The
Magic of
Mushrooms

DISCLAIMER

The information in this book is for informational purposes only and should not be relied upon as recommending, encouraging or promoting any specific diet or practice. It is also not intended as a guide to which mushrooms are edible or have nutritional or medicinal benefits or to replace the advice of a nutritionist, physician or medical practitioner. Specialist knowledge of nutrition, wild foods, fungi and plants is constantly evolving and you are encouraged to consult other sources and make independent judgments about the information discussed in this book. Neither the publisher, the author nor the Royal Botanic Gardens, Kew, are engaged or professionally trained in rendering nutritional, medical, or other professional advice or services.

There is the possibility of allergic or other adverse reactions from the use of any ingredients and foods and fungi and plants mentioned in this book. You should seek the advice of your doctor or other qualified health provider with any questions you may have, especially in relation to medical conditions or allergies. You should not use the information in this book as a substitute for medication or other treatment prescribed by your medical practitioner.

The authors, editors, publisher and the Royal Botanic Gardens, Kew, make no representations or warranties with respect to the accuracy, completeness, fitness for a particular purpose or currency of the contents of this book and exclude all liability to the extent permitted by law for any errors or omissions and for any loss damage or expense (whether direct or indirect) suffered by anyone relying on any information contained in this book.

Published in 2022 by Welbeck
An imprint of Welbeck Non-Fiction Limited,
part of Welbeck Publishing Group.

Based in London and Sydney
www.welbeckpublishing.com

The Royal Botanic Gardens, Kew logo and Kew images © The Board of Trustees of the Royal Botanic Gardens, Kew (Kew logo TM the Royal Botanic Gardens, Kew)

Text © Welbeck Non-fiction Limited 2022
Design © Welbeck Non-fiction Limited 2022

ISBN 978-1-78739-906-8

Printed in Dubai

10 9 8 7 6 5 4 3 2

Royal Botanic Gardens Kew

The
Magic of
Mushrooms

Fungi in folklore, superstition and traditional medicine

Sandra Lawrence

WELBECK

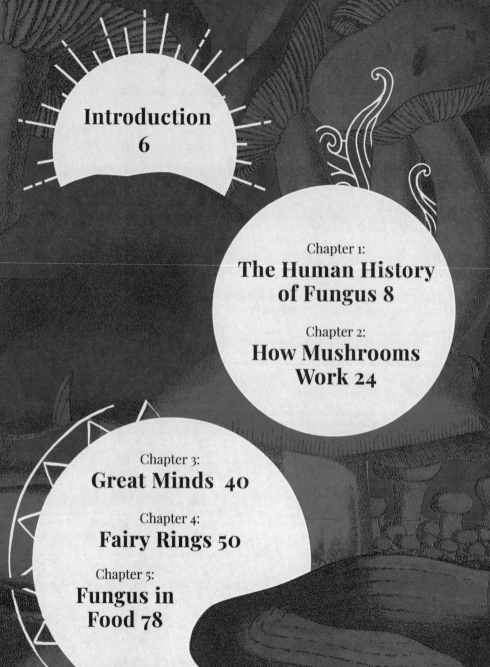

Introduction

Pale, fleshy, as if the decaying dead
With a spirit of growth had been animated
Percy Bysshe Shelley (1792–1822)

What is it about mushrooms that can both delight and horrify? Of all the kingdoms of life, only fungi seem to have consistently attracted a bad rap since antiquity. In the 1950s, Valentina and Robert Gordon Wasson divided the world into "mycophilic" and "mycophobic" countries: cultures where people are either comfortable with the idea of gathering and using mushrooms, or those where all fungi are "toadstools" and potentially poisonous.

Our ancestors tried to understand these strange life forms that could be alternately tasty or deadly – sometimes both in the same mushroom. Fungi had weird shapes, smelled unpleasant or appeared as if by magic. Legends, sayings and customs evolved to distinguish which was which, to explain how such entities came about and to indicate how they might be used as medicine, tinder, textiles and even, occasionally, food. Stories made sense of phenomena such as fairy rings, earthstars and, strangest of all, the fungi that appeared after thunderstorms.

For centuries, fungi were assumed to be unusual plant forms, which explains why the Royal Botanic Gardens, Kew, is home to the world's largest fungarium, cataloguing around 1.25 million specimens. After languishing for centuries as botany's poor relation, mycology is now a respected discipline, making discoveries that will change the world. Merely interpreting fungus on a scientific basis, however, misses something of vital importance: the way our often-uneasy relationship with it in the past may colour the way we regard this potentially life-saving kingdom in the future.

This book does not set out to explain fungus in a scientific way. There are already many volumes that do that much better than I could ever hope to; I've listed a few of my favourites in the bibliography. Neither is it intended as an identification or field guide. No one should try to work out what they've found in a forest by scanning a book of beautiful historical images and trying to fit their discovery to the most likely candidate. Instead, it takes a brief tour of the way fungi – and especially mushrooms – have infiltrated human existence. It traces a history of mythology and legend, superstition and suspicion, runs a thread through (a few of) the manifold uses of fungi in traditional medicines of the world and begins to look at how some of these practices are being investigated today.

Folklore rarely dies, but it does change. Indeed, many of our most colourfully named mushrooms have only been known as such for a few decades, after a new generation of romantically minded mycologists realized many spectacular specimens had no common names. Here, then, while gathering old sayings and stories, music and art, literature and poetry, I have also dipped a toe into modern interpretations within popular culture, where old themes die hard. Psychedelic pop art, slasher movies and computer games are really just modern skins on classic storytelling techniques.

Whether painted on the wall of an ancient Egyptian tomb or lurking in the ones and zeros of internet code, those stories largely remain the same: we are still fascinated, mystified and perhaps a little terrified by fungi, and we play out our ambivalence through myths and legends, old and new.

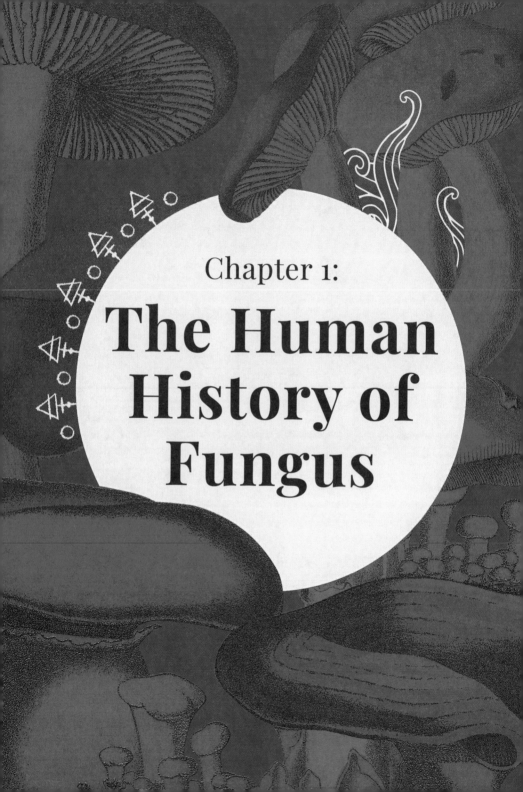

Chapter 1:

The Human History of Fungus

For most of our time on this planet, humans have been mystified and scared, yet still intrigued, by fungus. Every civilization has found ways of explaining its origins, while lending this potential friend or foe its due respect.

Around 18,700 years ago, a grieving community in Cantabria, northern Spain, buried a respected elder.

Daubed with red pigment and possibly garlanded with yellow flowers, the 35- to 40-year-old woman was important enough to merit her own grave marker in a side cave, now called El Mirón, close to the community's own living quarters.

We have no idea who the mysterious Red Lady is or why she was afforded such luxurious rites. But, thanks to modern dental analysis, we do know what she ate. Hers is the oldest known example of ingested fungus, though archaeologists can only guess whether it would have been consumed as food, flavouring or some kind of medicine.

Further north and later in time, Ötzi the Iceman had several uses for the mushrooms found in his bag, mummified alongside his body in the Ötztal Alps on the border of Italy and Austria, around 5,300 years ago. *Piptoporus betulinus* or *Fomitopsis betulina* (birch polypore) may have been a laxative to expel worms. Ötzi probably would have kept *Fomes fomentarius* (tinder fungus) for its fire-starting properties, though people have also used it to treat cancers, bladder complaints and even ingrowing toenails. It could be pressed into "mushroom felt" to make hats and small items, used to staunch wounds and, later, to light early firearms. The Ainu people of northern Japan traditionally burned the dried fungi to expel evil spirits during times of sickness.

In ancient Egypt, thanks to their curious property of appearing after thunderstorms, mushrooms were considered a gift from the storm god Set, and were sent to earth via lightning. Such treasures clearly conveyed immortality, so could be eaten only by the pharaoh. Common people were not even allowed to touch them, though they might admire certain temple pillars said to be in the shape of mushrooms: *Amanita*, perhaps, *Psilocybe*, or even various bracket fungi. If they ever caught sight of the pharaoh's crown, they might ponder that it looked a bit like a mushroom (though the similarity is not always agreed upon by scholars), but it's unlikely they would have witnessed the beautiful depictions of various mushrooms in papyri and tomb murals unless they were the artists who painted them.

The ancient Greeks also subscribed to the lightning theory, believing that mushrooms were plants with invisible seeds. This time it was Zeus who hurled the seeds to Earth during thunderstorms; mushrooms were therefore the "sons" of the gods. Aristotle classified fungus as a plant in his *Natural Philosophy* but he didn't really buy it. He couldn't help thinking they were something "other" and his puzzlement was shared by our ancestors. Without microscopes, they would not have been able to see airborne spores, and they would have been unaware of the vast networks of mycelia beneath their feet that give rise to ephemeral fruiting bodies that they would occasionally encounter. Fungal diseases may have been interpreted as malevolent divine intervention. Mushrooms were considered good to eat, though risky – cases of accidental (and

Opposite *Amanita* mushrooms from J.V. Krombholz, *Naturgetreue abbildungen und beschreibungen der essbaren, schädlichen und verdächtigen schwämme,* 1831–46.

1 5 Aſchgrauer Fliegenſchwamm AMANITA CINEREA, Otto. 6 9 Feinfilziger brau-
ßantherflechiger Fliegenſchwamm AMANITA PANTHERINA, D. C. Franz. Golmotte fautse
8 11 Schuppiger Fliegenſchwamm AMANITA ASPERA, Pers. Franz Agaric âpre.

Above: *Two cranes with the red spot of divineness on their head under pines beside the mushroom of immortality* by Lu Ji. Indian ink and paint on silk, *c.* 1520.

otherwise) poisoning were common. Farming them seemed a much better bet, and Greek biologist Theophrastus suggested that mushrooms grown on dunghills were not necessarily evil.

One fungus, agarikon (*Fomitopsis officinalis*) was the ancient world's medical get-out-of-jail-free card: a panacea, a cure-all, that was still being recommended by John Gerard in his herbal of 1597 to help ease, among other things, asthma, poor digestion, worms and the coughing of blood from the lungs. Known as the "bread of ghosts" by the indigenous Tsimshian people of North America, *Fomitopsis officinalis* still carries spiritual and medicinal importance for many communities today.

The Romans liked *Fomitopsis officinalis* too, but they were much more interested in fungi as food. Even though calling someone a "toadstool" was a prime insult in ancient Rome, mushrooms were celebrated – even guzzled – by anyone allowed to eat them, which didn't include the common folk. By law, only nobles were permitted to eat mushrooms, which were cooked in special pots called *boletaria*. They could get quite piggish over *Boletus edulis*, or "little pigs" (porcini).

Warriors were also allowed to eat mushrooms that, ingested immediately before battle, would improve their strength and endurance. The soldier-scientist Pliny the Elder wrote of all mushrooms under the umbrella name "agarikon", so it's hard to tell which species he is referring to in his discussions. There was a generally distinct cooling of scientific ardour for fungus during this time, though. Plutarch and Pliny, like Aristotle, didn't understand fungus. They had trouble classifying it and eventually began to lose interest. The first-century physician Dioscorides had similar problems writing his seminal herbal, *De Materia Medica*. He, like others, believed that mushrooms were made out of decaying damp earth, were poisonous – or at least indigestible – and had no nutritional value as food.

Since *De Materia Medica* would be the go-to Western medical work for 1,500 years, fungus slowly became "unfashionable" in Europe, save for poorer people who would have been only too aware of – and grateful for – the various edible forms. That did not mean the rest of the world was ignoring mushrooms. In China and Japan, they were considered life-extending, even lenders of immortality. The mythical "Sage King" Shennong is considered the father of traditional Chinese medicine (TCM). When he wasn't inventing the plough, hoe and axe, or teaching people farming techniques, he is said to have personally tested herbal drugs on himself. The *Shennong Ben Cao Jing* (*Divine Farmer's Classic of Pharmaceutics*), written around 200 to 250 CE, classifies various mushrooms as aids to the life forces of different parts of the body. It notes that fungi are to be carefully regulated in order to extend human life.

The curious "mushroom stones" of Mesoamerica were considered to be offensive pagan idols by later Catholic missionaries and vast numbers of them were destroyed, but enough survive for us to peer into a mysterious world. Evidence for this mushroom cult, which died out centuries before the missionaries arrived, is found mainly in Mexico and Guatemala. Strange, phallic mushrooms carved from stone and usually growing from the backs of kneeling figures may imply a worship of hallucinogenic fungi. In truth, no one really knows what ancient worshippers believed. Later, the Aztecs called mushrooms "flesh of the gods", saying that the great god Quetzalcoatl had formed them from drops of his own blood.

In a European world suspicious of fungus, the twelfth-century German saint, herbalist, philosopher,

mystic and musician Hildegard of Bingen stands out as exceptional. Believing all things on Earth were sent from God, she was curious about what uses fungi might have, and decided that all mushrooms that grow on trees must be either edible or medicinal. Other Christian texts were less generous. Certain scholars, including John M. Allegro, have suggested a tradition in medieval illumination that depicts trees as mushrooms. The biblical tree of knowledge is often, such scholars suggest, portrayed as a giant *Psilocybe* mushroom; the serpent winds itself around its stipe (stem), eyed by a sceptical Eve.

Eastern cultures, including Slavic countries such as Russia, Poland and the Czech Republic, as well as Hungary, had fewer hang-ups, enjoying fungus as both food and medicine.

Italian botanist Pier Antonio Micheli's *Nova plantarum genera* (*New genera of plants*), written in 1729 but never published, saw the fungal logjam begin to shift as he described no fewer than 900 fungi and lichens never before scientifically observed.

Many of Micheli's contemporaries were unconvinced by his suggestion that fungi have reproductive bodies, but he stuck to his guns, proving his theory by creating cultures from spores and essentially giving birth to mycology – the study of fungus. His 73 plates of various fungi sparked new interest in what had been a murky, poisonous world of sinister entities. Twenty years later, fungi came out of the folkloric closet into the dazzling light of scientific observation. Swedish "father of taxonomy" Carl Linnaeus included fungi in his new "binomial" (two-word) system of classification, though he, too, considered fungus part of the plant kingdom.

The word "mycology" was first used by Miles Joseph Berkeley, an English clergyman and fanatical fungus collector. Farmers had particular reason to

Mushroom Monstrosities —

thank Berkeley, as he pinpointed specific fungal pathogens on various food crops. Knowing one's enemy is the first step to fighting them, and his investigations into the causes of the Irish potato famine of the 1840s led to the culprit – *Phytophthora infestans* – being identified. It is now classified as an oomycete, more closely related to brown algae than any fungus.

The nineteenth century saw an explosion of interest in mycology, often from female scientists. While pursuing science was not generally considered proper for a nice young lady, botany was permitted. Several women, such as Anna Maria Hussey and Gulielma Lister, whose personal passion was for

Opposite Chicken of the woods (*Laetiporus sulphureus*) from Anna Maria Hussey *Illustrations of British Mycology*, 1847–55.

Above *Mushroom Monstrosities* humorous print by George Cruikshank, 1835.

slime mould, led the charge. The Victorians were fascinated by all things nature-related, yet that did not stop one of Britain's most famous mycologists – Beatrix Potter – becoming more famous for children's stories than her hundreds of detailed botanical illustrations of fungi and lichens.

By the twentieth century, mycology had developed into a discipline that seemed to have as many branches as there are divisions in a mycorrhizal network. The discovery of penicillin by Scottish microbiologist Alexander Fleming in 1928 proved that folk traditions of using mouldy bread to cure infected wounds might have something in them after all. Hildegard of Bingen is rarely cited today, and yet our modern impetus to explore the medical possibilities of the planet's organisms echoes the spiritual joy she found in the interrelationship between humans and the natural world. We have made much progress, but are still nowhere near understanding this most mysterious of the earthly kingdoms.

Caesar's mushroom

Amanita caesarea

Amanita caesarea is named for the Emperor
Claudius (10 BCE–45 CE), who adored the fungus
to the point of gluttony.

Native to Southern Europe and Northern Africa, this smooth, rounded mushroom, with its bright orange cap and creamy yellow gills, was hugely enjoyed by the nobility and the army. North of the Alps, *Amanita caesarea* is often found along old Roman roads where it is said centurions on the march discarded their leftovers. It has been found as far as Eastern Europe, India, China and even Mexico. Caesar's mushroom is a member of the *Amanita* genus, related to *Amanita muscaria* (fly agaric) and depicted in murals and mosaics.

Three different historical sources – Tacitus, Dio Cassius and the entertaining if unreliable Suetonius – tell us that in 54 CE, the ageing Claudius died from a dish of mushrooms laced with poison. It was rumoured his fourth wife, Agrippina, was behind the deed, determined that Nero, her son from a previous marriage, should succeed his stepfather, rather than Claudius's own son Brittanicus.

Agrippina reportedly employed a poison-mistress, Locusta, whose brief was specific. Claudius must walk away from the feast seemingly unharmed. The effects, a few hours later, must be devastating and deprive the victim of his faculties, in case he realized what was happening and tried to punish the perpetrator. One poison available at the time fits the bill – *Amanita phalloides*, the death cap. Some historians are sceptical that Locusta, or any Roman, could have understood the precise way death cap can kill, and none of the historical sources relating to the murder were contemporary with the poisoning. However, one writer is remarkably quiet on the subject. The philosopher, politician and playwright Seneca was not only Nero's tutor but also may have been witness to the event, yet he makes only satirical allusions to the mystery. In 1972, ethnomycologist Robert Gordon Wasson found an odd phrase in one of Seneca's letters, proving the statesman knew all about the effects of *Amanita phalloides*. Wasson believes Seneca implies that, rather than a mere lacing of poison, Claudius was served a whole dish of death caps while everyone else got regular mushrooms.

After his death, Claudius was declared a god. Suetonius relates how, years later, someone told Nero that mushrooms were the "food of the gods". Nero apparently replied, "True enough, my father was made a god by eating a mushroom."

Opposite *Amanita caesarea* from JL Leveille *Iconographie des Champignons de Paulet*, 1855.

Fig.1.2.3. M .... Hypophyllum cæsareum

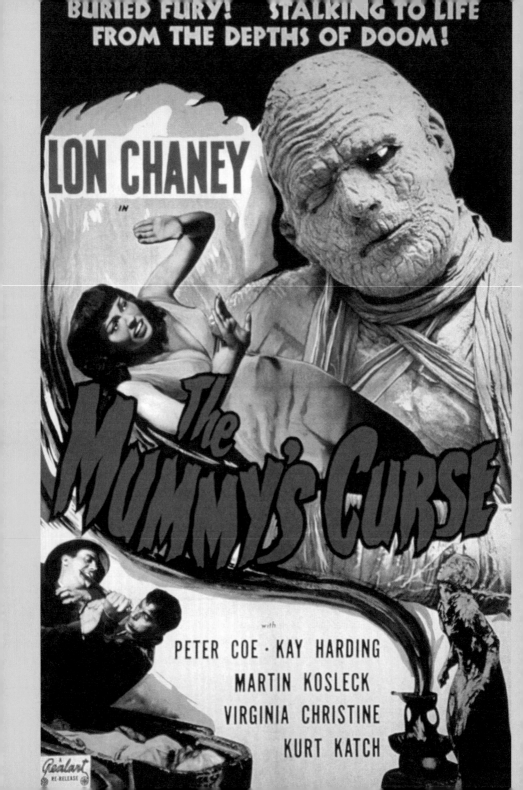

The mummy's curse

Aspergillus Species

*Death shall come on swift wings to him
who disturbs the peace of the king.*

These words, rumoured to be written on a tablet at the entrance of Tutankhamun's tomb, predicted doom. As archaeologists dug in November 1922, they were circled by vultures, the symbol of Lower Egypt. A cobra, the sign of Upper Egypt, devoured Howard Carter's pet canary, even as the archaeologist tempted fate and broke the tomb's seal. Four months later, the expedition's patron Lord Carnarvon fell gravely ill in Cairo. As the earl wheezed his last, the city's electricity failed. Back in England, his faithful dog howled and dropped dead. By 1929, 11 people connected with the discovery had died strange deaths.

The press went wild, theorizing that Carnarvon's death had been caused by "elementals". Hollywood was obsessed with the idea of a "mummy's curse", a fascination that has never waned. That was exactly what the ancient Egyptians wanted – robbers must be dissuaded at any cost. The written "curse" was a newspaper invention, but the idea was sound.

Howard Carter didn't believe a word of it, but he wasn't daft. His famous reply to Carnarvon's "Can you see anything?" was "Yes, wonderful things," but Carter was concerned enough about ancient fungal spores to take specimens of tomb air and samples from the sarcophagus before going any further. They were clean, so he continued his work. The following deaths were diverse enough not to be related, but fuel rumours of a curse to this day.

In Poland, 1973, the recently discovered tomb of King Casimir IV saw four researchers die within four days and six more soon after. *Aspergillus flavus* was found in the tomb, but no autopsies were conducted, and modern scientists are sceptical that it could have had such a strong effect on so many apparently healthy people. In 1999, microbiologist Gotthard Kramer studied 40 mummies and identified several species of mould in Egyptian tombs. He suggested that when tombs were opened, fresh air blew the microbes into the open, infecting the archaeologists' ears, noses and eyes. Other microbiologists found three types of *Aspergillus* spores at Egyptian archaeological sites and claimed infections suffered by workers were compatible with *Aspergillus*.

Today, the mummy's curse appears to be working in reverse, as the breath of millions of tourists brings dampness and spores to eat at the walls of the royal tombs. As the murals blacken and crumble, the only answer seems to be to permanently seal the tombs again.

Opposite Poster for *The Mummy's Curse* (1944).

Lingzhi

Ganoderma lucidum

The 1596 Chinese classic *Bencao Gangmu* (*Compendium of Materia Medica*) identified six coloured fungi as the "herbs of immortality".

Each had a different beneficial effect and, taken regularly, claimed author Li Shizhen, could lighten one's weight, positively affect life energy and extend the lifespan to that of "the immortal fairies".

The word "lingzhi" first appeared in Zhang Heng's poem "Xijing Fu" sometime in the first century CE, but had not yet been narrowed down to a single species. Today, most people equate the "divine mushroom" with one fungus: lingzhi (*Ganoderma lucidum*).

It is in its dozens of folk names that this mushroom truly comes to life: "spirit plant", "tree-of-life mushroom", "forest sage", "10,000-year mushroom". In Japan, the reishi is known as the "good fortune mushroom", "varnished conk" and "monkey's seat". *Ganoderma lucidum* (from the Latin *gan* for "shiny", *derma* for "skin") is a woody bracket fungus, sometimes called the "lacquered bracket" for its glossy surface, that grows from the base of mature trees.

There's no evidence that *Ganoderma lucidum* has been used, medicinally or spiritually, for 4,000 years, but the legends surrounding a mushroom of immortality are so powerful it is impossible to separate myth from history. Certain traditions associate it with the mother goddess and Guanyin,

the goddess of healing, is also occasionally depicted holding lingzhi. In some versions of the *Legend of the White Snake*, one of China's four great folk tales, the magical herb that Madame White Snake steals to resurrect her dead husband takes mushroom form.

Such tales are, unsurprisingly, associated with the human quest for longevity. Legend holds that long before he declared himself the first emperor of China, Qin Shi Huang – a real historical figure of the Qin dynasty (221–206 BCE) – met a 1,000-year-old wizard who took him to the mystical Isles of the Blessed. Hidden in a bottomless abyss, these islands float in the air but vanish on sight to all but wizards. They are verdant with magical trees, with jewels for fruit, stone-shaped mushrooms and curious creatures, and are inhabited by transparent spirits: saintly Taoist sages who have won immortality and shed their human bodies. These spirits renew their youth daily by eating the fungus of immortality, which lends them the power to float between the islands.

On returning to his world, Qin Shi Huang became one of many emperors obsessed with finding the elixir

Opposite *Ganoderma lucidum* from James Sowerby *Coloured Figures of English Fungi or Mushrooms*, 1795–1815.

Jan.ᵗ 1790. Publish'd by J. Sowerby London.

Boletus lucidus

of immortality. After three unsuccessful expeditions, he sent the alchemist-explorer Xu Fu, together with a battalion of ships filled with children, to continue the quest. None returned. Some say Qin Shi Huang died testing a potential elixir, and was buried with his entire court inside a pyramid flowing with underground rivers of mercury. The story is legend, but archaeologists have found concentrated levels of mercury near the site of the First Emperor's Terracotta Army.

Unfortunately, several elixirs the emperors drank while waiting for the fungus of immortality contained mercury, arsenic and other poisons. Some scholars have suggested that the quest for the fungus actually sought an antidote to elixir poisoning.

A similar story talks of Han dynasty courtier Dongfang Shuo (c.160–93 BCE), whose search for the fruits of life, including the famous fungus, was probably exaggerated even when his tall tales were first told. Even during his lifetime, Dongfang sold himself as an immortal. His stories of the floating, vanishing Isles of the Fortunate, where the fruit of immortality grew, clearly built on older traditions.

None of these may have been the lingzhi that bears the title today. Early twentieth-century anthropologist Berthold Laufer suggested the real fungus of immortality was a species of agaric, "a felicitous plant

because it absorbs the vapours of the earth". *Huang Di Nei Jing* (*The Yellow Emperor's Classic of Medicine*, c. 300 BCE) refers to lingzhi as "tiny excrescences" – a concoction of magical substances including minerals, fossils, gems, lichens and fungi. Taken long-term, lingzhi could lend the "intermediate class" several thousand years. Even lower classes could expect a millennial lifespan. The claim was theoretical – such food was forbidden to anyone but the emperor.

Being forbidden gave lingzhi an air of mystery, and kept prices buoyant. Today, *Ganoderma lucidum* is prized in traditional Chinese medicine, where it enjoys a status almost as high as ginseng. Considered warming, astringent, nourishing and detoxifying, it has been used for ailments including cancers, heart, kidney and bronchial problems, hepatitis, nerve pain and insomnia. This complex range is not dismissed by conventional Western medicine, and lingzhi has been the subject of thousands of recent studies probing possible therapeutic uses.

Ganoderma lucidum is most prized in its wild, whole form, sold dried or fresh in markets. But most lingzhi today is commercially grown in sterile conditions, in a reported $2.5 billion business. Anyone today may enjoy immortality – as a tablet, food supplement or even in alcoholic drinks.

Opposite Ming dynasty silk tapestry of Dongfang Shuo stealing a peach of immortality, *c.* 1368–1644.

Right Scene from the *Legend of the White Snake*. China, late nineteenth–early twentieth century.

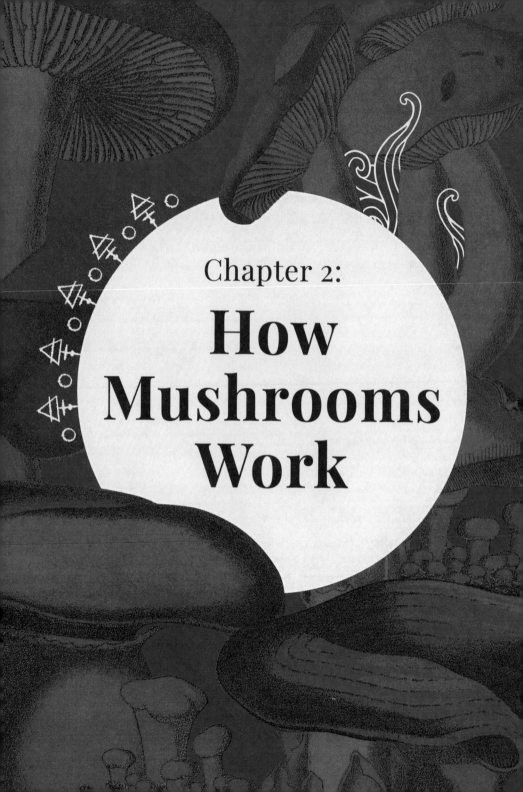

Chapter 2:

How Mushrooms Work

For millennia, scientists found fungi a complete enigma. Mushrooms grew out of the ground like plants but, unlike them, were unable to make their own food via the sun. They couldn't be animals, either, because they didn't move, and though some suggested mushrooms might be mineral "excrescences" of the earth, few were convinced that stones were ever once alive. Fungi were squeezed into a corner of the plant kingdom and only awarded their own independent lineage in 1968.

Robert Whittaker (1920–1980) was the first scientist to be taken seriously in proposing fungi as one of the five kingdoms of life, but he wasn't the first to think of it.

Few had thought the idea worth adopting. A century earlier, Ernst Haeckel (1834–1919) had proposed a three-kingdom system of animals (Animalia), plants (Plantae) and "everything else" (Protista), including fungi, protozoa, bacteria and other microorganisms. Whittaker's 1968 five-kingdom theory was widely accepted – until the advent of DNA, which is still shifting our concepts of what "belongs" where. Opinions differ as to how many kingdoms life on Earth currently has, but no one denies that fungi are out there on their own.

Even so, it is a broad classification, including yeasts, moulds, mildews and mushrooms. The earliest fungi are thought to have been simple, single-cell organisms, living in water as part of the "primordial soup", around a billion years ago. They moved around using a flagellum, a whip-like structure, and reproduced via asexual spores (zoospores). Fungi are believed to have made it onto land around 700 million years ago, but research is moving fast and new discoveries are being made all the time. It is currently estimated there are between 2.2 and 3.8 million species of fungi; so far humans have identified fewer than five percent of them.

To many people's surprise, fungi have more in common with animals than plants. Plant cell walls are formed of cellulose; whereas in fungi they are made of chitin, the same substance that forms the exoskeletons of insects and Crustacea. Indeed, chitin is not dissimilar to the keratin found in human hair and skin. Fungi store their food as glycogen, in a similar way to animals, whereas plants store their nutrients as starch. Unlike plants, which are autotrophs and are therefore capable of producing energy through photosynthesis, fungi are heterotrophs just like animals. This means that they gain their nutrition from other organisms, and fungi go about this in a diverse range of ways. Saprotrophs, such as turkey tail (*Trametes*) and oyster mushrooms (*Pleurotus*) decompose dead matter, breaking down cells to get to the nutrients. Unlike animals, which move towards their food and use some sort of mouth to ingest their nutrition, which is then digested in an internal stomach containing enzymes, saprotrophic fungi can grow in and around their food source, releasing enzymes outside of their cells that break down organic matter externally, which is then reabsorbed through their cell walls as nutrition. Sometimes they "speculate" by latently lurking in a diseased or dying plant, waiting for their host's demise. They can be incredibly destructive, as in the case of dry rot (*Serpula lacrymans*) but need certain conditions – including oxygen, water, neutral acidic conditions and a low temperature – to thrive. This is partly why archaeological finds in anaerobic (oxygen-free) conditions – for example the wreck of the sixteenth-century warship *Mary Rose* – sometimes have not decomposed. Yet this process of decomposition, of which the fungi are the great masters, is essential to all life on earth and has created the rich soils upon which we all rely.

Above Twelfth-century fresco of Adam and Eve –
the Tree of Life resembles a giant mushroom.

Biotrophic fungi create a long-term relationship with other organisms to gain their nutrition. Some deplete their hosts' resources, others work in ways that are mutually beneficial to both parties, connecting entire ecosystems. Necrotrophs kill their host plant and asset-strip the nutrients.

Most of the time, fungi live in darkness, out of sight, often in the ground or inside plants. They grow and operate via the mycelium, a network of thin, branching tubes (or hyphae) that carry water and nutrients through the soil. Unlike animal cells, which have just one nucleus, many fungi are polykaryotic (having multiple nuclei) and multicellular, with special pores between each cell that allow the cytoplasm (a thick, solution of water, salts and proteins) to move freely between various parts of the organism.

These multicellular threads can act as mycorrhizas, a system that connects the fungi with plant roots in a mutually symbiotic relationship in which the tree exchanges carbohydrates produced by photosynthesis for water and nutrients the fungi forages from the forest floor. Mycelium can spread underground for hundreds of miles, acting as a communication system that may or may not include other organisms. Mycelium bonds the soil, cleans away debris, distributes nutrients and, in its turn, becomes food itself.

Humans most often associate the idea of a fungus with an image of its reproductive organs. The fruiting body of a fungus is only a tiny part of the whole, but it is vital in order to produce spores which, when dispersed, can spawn offspring. While many fungi – such as yeasts – merely divide, some are capable of reproducing both asexually and sexually. The "sexual encounter", which consists of the fusing of two nuclei, each with its own chromosomes (thread-like packages of DNA), can only be witnessed under a microscope. The resulting cells produce spores, the basis of new fungal life. Microscopic scrutiny will also reveal that these miniscule reproductive cells are never just boring dots. Ranging in size from 0.003mm long to specks large enough to be visible by the naked eye, the sheer diversity of fungal spores is extraordinary. Threads, coils, branches, blobs, and even shapes that resemble seashells, stars and marine creatures. They sometimes also have a jellylike sheath, or appendages not unlike heads or tails. Some have several walls and/or compartments, or an outer shell; others are slimy, dry, projectile or smooth, designed merely to fall to the ground. They can be any hue, from transparent to black, pink, brown and purple; some are even multi-coloured.

The two largest groups of fungus, Ascomycota and Basidiomycota conceal their spore-producing cells in special distribution structures. Ascomycetes form sacs called asci, which are then distributed via any number of methods, from discharging like a pistol to merely breaking down into a mass of spores. Basidiomycetes produce their spores on the tips of basidia (club-shaped structures) that develop on certain parts of a fungus, ready for dispersal by the wind or grazing invertebrates. These are often present in mushrooms as gills: folds or "leaves" below the cap that protect the spores until they are ripe and environmental conditions are at their most advantageous for germination.

It is these "mushrooms" that have fascinated, thrilled and, on occasion, horrified humans into making meals, hurling curses, finding magic and telling tales of them since time immemorial.

Opposite Turkey tail (*Trametes versicolor*) from Anna Maria Hussey *Illustrations of British Mycology*, 1847–55.

Naming of Parts – The Anatomy of a Cap Mushroom

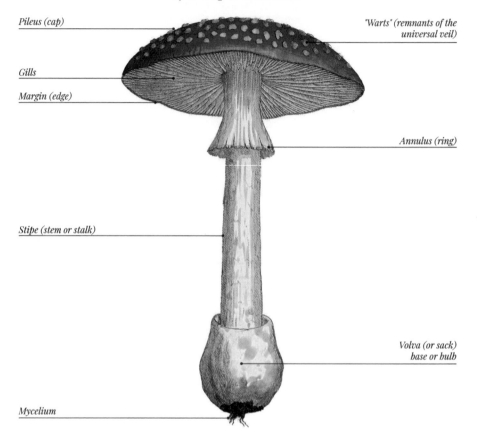

Pileus (cap)

"Warts" (remnants of the universal veil)

Gills

Margin (edge)

Annulus (ring)

Stipe (stem or stalk)

Volva (or sack) base or bulb

Mycelium

For centuries, people believed that the fruiting body was the "main" part of the fungus, so most folklore surrounds the "visible" mushrooms. They are as different as they are numerous, but many share similar properties.

Mycelium

Traditionally regarded as the mushroom's "roots", mycelium, we now know, form every part of the fungal organism. Branching networks of hyphae (thread-like tubes) can stretch (usually underground) for hundreds of miles. Mycelium transport nutrients, can sense and move towards food sources or symbiotic partners, and can even be used to grapple and attack competitor fungi. Even the fruitbodies are made of these hyphal cells, which adapt to form different structures, eventually giving rise to new mycelia and fruitbodies of their own.

Pileus (cap)

The cap can be many shapes, ranging from extremely concave to extremely convex, almost like a cone, and take on many colours, which may change as the mushroom matures.

Stipe (stem or stalk)

Not all mushrooms are "stipitate" (fungi with stipes), using a stalk to support the cap and gills, lifting them above the ground for protection and more effective distribution of the spores. Like the rest of the fruiting body, the stipe consists of sterile hyphae. It can be hollow or solid, fibrous or soft; none of these qualities will tell a forager whether or not the mushroom is poisonous.

Universal veil

Immature gilled mushrooms will often be covered by a membrane of tissue, enclosing the whole fruitbody in its immature state. This will split or fracture as the fruiting body grows larger and, depending on its cellular structure, can leave a range of different structures and patterns on the mature fruitbody. In the case of the fly agaric, the universal veil is made up of powdery warts that initially are entirely conjoined, forming a spiny egglike cover. As the mushroom swells and extends, the universal veil breaks up into small isolated warts, some of which stick to the base of the stipe while others remain on the cap, giving this fungus its spotty appearance.

Volva

Also sometimes known as the sack or bulb, the volva is the remnants of a universal veil and is very rubbery or membranous. The universal veil doesn't break into break into smaller pieces but instead tears open, allowing the maturing mushroom to burst through, leaving a saggy sacklike structure at the base, known as a volva.

Annulus (ring or partial veil)

Found half- to three-quarters of the way up the stipe, the annulus is found on mushrooms that have partial veils (a disc of tissue that connects the stipe to the cap margin). The ring is what is left after the veil has split away to reveal the gills. Depending on the fruiting body, the annulus can be large and skirtlike, flare out like a ballerina's tutu or act more like a sheath, following the lines of the stipe. It may look like delicate lace, become translucent and like a cobweb, or be solid, thick and fibrous.

Partial veil

A thin membrane on some mushrooms that covers the gills until the fungus is ready to distribute the spores.

Lamella (gills)

The gills are thin "ribs" that radiate from the stipe on the underside of some mushrooms. They act as a means of spore dispersal, protecting the spores in their folds, which also lend extra surface area for the spore-producing cells. They can be "crowded" below the cap or widely spaced, appear as single entities or look like folds, ribs or veins. Some drip latex. Gills are not necessarily the same colour as either the cap or spores.

Spores

A mushroom's spores contain everything it needs to replicate itself. Spores may fall from the gills or rely on other creatures, like insects, for distribution. The lightest spores are puffed into the air to be borne away via air currents.

From A to B – Ascomycota to Basodiomycota and Beyond

Of the many major groups (phyla) of fungi, most are invisible: soil-dwelling "rotters" that digest detritus or microscopic parasites or fungi that live in the guts of plant-eaters like sheep. Of the rest, two phyla particularly resonate with humans, but in the fungus kingdom there are always oddballs.

Commonly found members of the Ascomycota usually have extraordinary, strange-looking fruiting bodies. The phylum includes truffles and morels, which have developed profound aromas to attract the animals that will distribute the spores – but just to keep us on our toes, it also includes *Penicillium*, a green mould and our first antibiotic. Basidiomycota contains many diverse groups, but we are most familiar with what we have traditionally called mushrooms, but even these can be broken down into different types of fruiting body.

"Mushrooms"

A gigantic group of fungi, these Basidiomycetes (from the Basidiomycota phylum) take on hundreds of different guises in every colour of the rainbow. They typically have many basidia (literally "little pedestals") on their gills, from which their spores are produced. Members of the Agaricales order usually have a fleshy sporing body with a cap, gills and a stem. They are found across all continents, and although usually preferring grassland and woodland, can be found in many places – even, in one case, fruiting underwater.

Psathyrella aquatica, found in the Rogue River, Oregon, in 2005 by Professor Robert Coffan, is currently considered unique, though as with all discoveries in the fungal kingdom, it may yet turn out to have relatives.

Waxcaps (*Hygrocybe* spp. and other related genera) are some of the brightest-coloured of the Agaricales, appearing in jewel-bright shades

Below Bog beacon (*Mitrula paludosa*), an ascomycete, from Anna Maria Hussey *Illustrations of British Mycology*, 1847–55.

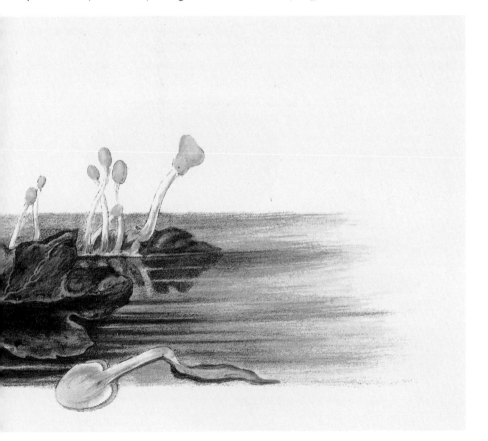

ranging from deep scarlet, shocking pink, sunshine yellow and dark orange to pastel rose, snowy white, lime green and deep black. They thrive on ancient European grassland, but also appear in undisturbed pasture, coastal walks and long-established lawns – for example, in the grounds of stately homes. In North America, they are often found in woodlands. Many are considered endangered species since once the ground is disturbed it can take decades or even centuries for waxcaps to re-establish. Other examples include *Coprinus comatus* (shaggy ink cap), *Stropharia caerulea* (blue roundhead), *Cortinarius violaceus* (violet webcap) and *Amanita muscaria* (fly agaric).

Bracket fungi (conks)

Although markedly different in appearance from "regular" mushrooms, polypores or bracket fungi are also basidiomycetes. They usually grow out from tree trunks or branches, supported by fleshy growth that looks a bit like shelf brackets. Most polypores live inside the tree, consuming the non-functional heartwood in the interior; the visible part is merely the fruiting body, though a few live in the soil, forming mycorrhizae with trees. The fleshy underside of polypores is made up of thousands of tiny tubes or pores, rather than the gills produced by mushrooms. Bracket fungi are hugely important to the planet, as they are the only organisms that can break down the lignin – tough, woody compounds – in trees. By decomposing the lignin, the fungus makes nutrients previously locked up available for other organisms to feed on. We see it as "rot", but without it, the planet

would be littered with dead matter. The unfortunate flip side of this is that the fungus sometimes breaks down wood that trees – or humans – are still using, occasionally killing trees and damaging timber. Some bracket fungi will only grow in very old woodland, and are used to identify ancient forests. Their uses to humans are many and varied, from tinder to hats and from food to medicine. *Inonotus obliquus* (chaga), for example, has been found to suppress cancer-cell growth and enhance human immune systems.

Examples include *Laetiporus sulphureus* (chicken-of-the-woods), *Buglossoporus quercinus* (oak polypore), *Trametes versicolor* (turkey tail) and *Cerioporus squamosus* (dryad's saddle).

Cup fungi

These fungi produce sporing bodies that are shaped like little cups, called "apothecia" and are usually ascomycetes (from the Ascomycota phylum). Their tiny "cups" house the spores until ready for dispersal, when most "shoot" their cargo into the atmosphere. The weather conditions need to be right, but when they are, the tiniest puff of air will disperse the spores. Cup fungi can be found in all environments – animal dung, sandy beaches, dead plants and dank woodlands – but many are fleshy and soft and react badly to being dried out. Most live on dead matter, but some create mycorrhizal relationships or dwell inside living plants as endophytes.

Examples include *Sarcoscypha austriaca* (scarlet elf cup), *Otidea onotica* (hare's ear), *Trichoglossum hirsutum* (black earth tongue) and *Scutellinia scutellata* (eyelash cup).

Opposite Violet webcap (*Cortinarius violaceus*), a basidiomycete, from Anna Maria Hussey *Illustrations of British Mycology*, 1847–55.

Gasteromycetes

Gasteromycetes is a group of fungi with similar reproductive strategies, even though they do not all share the same family, or even a phylum. One thing they do have in common is the inability to forcibly disperse their spores from gills or pores. Instead, they have evolved ingenious alternatives that make them some of nature's most beautiful – and strange – shapes. The simplest are puffballs: round, often pure-white vessels that can range in size from a few millimetres to well over one-and-a-half metres across. Some have small holes through which they "puff" clouds of spores on impact – perhaps from a passing animal or even a single raindrop. Others split open when they are ripe.

Bird's nest fungi are shaped like cups filled with little "eggs" that contain spores. When a rain drop lands in the cup these "eggs" are launched out into the environment, sticking to nearby plants and eventually opening to release the spores. Other fungi make themselves attractive to animals, which disperse the spores through their droppings. Perhaps strangest of all are the stinkhorn fungi (*Phallus* spp.), which produce a foul-smelling (to humans) spore-laden liquid called a gleba, which the fungus presents on the tip of its sporing body. Flies that usually feed on faecal or dead matter are attracted to the gleba, often taking mere minutes to devour the lot. As the fly buzzes off to find its next meal, all six of its feet are covered in sticky spores, which the insect distributes without even knowing the lunch was not free.

Examples include: *Lycoperdon perlatum* (common puffball), *Cyathus striatus* (bird's nest), *Geastrum quadrifidum* (earthstar), *Colus hirudinosus* (red cage or lattice fungus) and *Phallus indusiatus* (veiled lady).

Lichen

Lichens are some of the oldest organisms on the planet, thought to have evolved around 450 million years ago, the same time as fish. However, they continue to evade complete understanding.

For a lichenologist, finding a new species, especially somewhere as unexpected as the British Isles, is a possibility, though it is getting harder. The reason is simple: pollution. Lichens have long been considered bellwethers for environmental problems. Though many are tough, others must have clean air; without it, they are the first to die.

Lichens transcend kingdoms, being a combination of fungi and algae and/or cyanobacteria living in symbiosis. The fungus provides a protective environment for its photobiont – an alga or cyanobacterium – which photosynthesizes sugars for both. The relationship has been likened to farming because the fungus cultivates its photobiont. However, lichen is epiphytic: it grows on its host without being parasitic, though it can harvest minerals. It is just as happy using a dead tree, rock, roof – even an unwashed car – as its anchor .

They often appear as low-lying "crusts", in a rainbow of colours, on the bark of trees or rock faces, but can also be almost bushy in appearance, creating an "old man's beard" of tendrils and branches of their own. *Usnea* has many names worldwide, including "Merlin's beard", "witches' whiskers", "fishbone beard", "tree's dandruff" and "seaweed of the mountain". In China, *Usnea diffracta* is known as "Lao Tzu's beard", after a semi-legendary philosopher (c. fourth to sixth centuries BCE) who was born fully grown with a beard, after his mother's 62-year pregnancy.

Lichen grows across the world, but one of the most diverse populations is in Scotland, thanks to a range of perfect-condition habitats – high humidity, sunless days, persistent and low clouds, temperate climate and sparse human population. Lichens flourish by Scottish seas, lochs, woodlands, peat bogs and mountains, in colours that have not gone unnoticed by the locals there and in similar lichen havens, such as Shetland, the Hebrides and Wales. Lichens have been scraped from the rocks by crofters for centuries, and boiled in three-footed cauldrons into a range of dyes still used today. The most famous use is in Harris Tweed, as a blend of tints incorporating the soft tones of Scottish lichens. Dartmoor, in England, also has a strong tradition of lichen dyes. Another lonely place, the moor has been largely left to its natural inhabitants for millennia.

Opposite Beard lichen (*Usnea hirta*) and witches' beard (*U. florida*) from G.F. Hoffmann *Descriptio et adumbratio plantarum e classe cryptogamica Linnaei quae lichenes dicuntur*, 1790–1801.

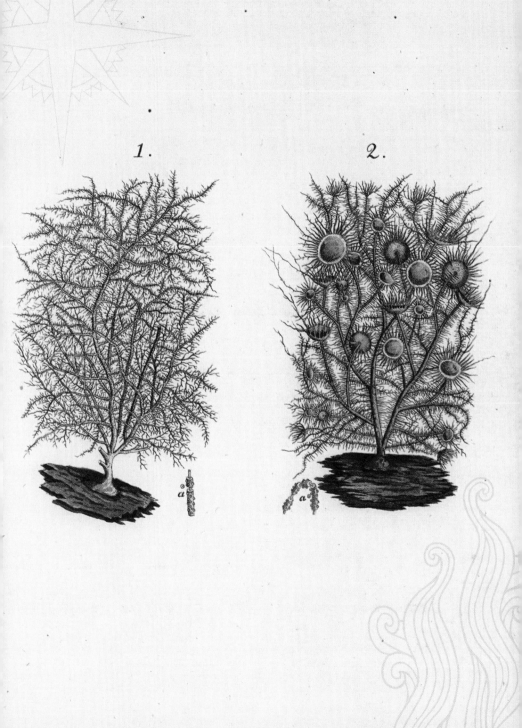

1.

2.

a

a

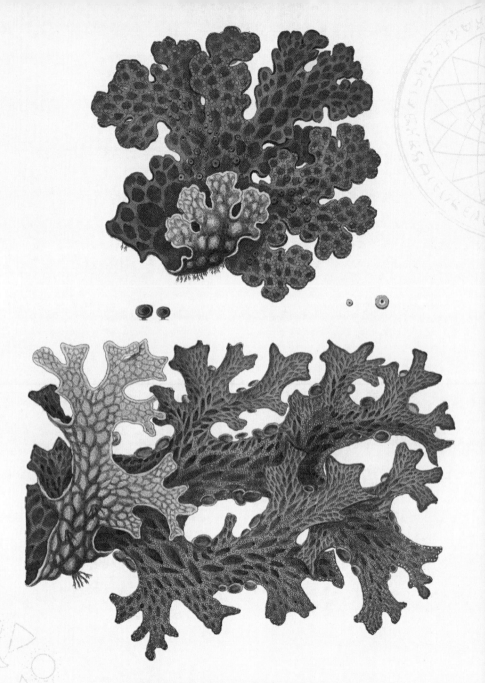

Two lungwort lichens (*Lobarina scrobiculata* and *Lobaria pulmonaria*) from G.F. Hoffmann *Descriptio et adumbratio plantarum e classe cryptogamica Linnaei quae lichenes dicuntur*, 1790–1801.

Some have suggested lichen was the first "tinsel" used on Christmas trees, but it has been many things to many people. To the Gitxsan people of Canada, *Lobaria* (lungwort) was useful in rituals for bringing health and longevity; other tribes may have seen lichens as aphrodisiacs or ways to become "invisible". Commercial uses for lichen include toothpaste and perfume.

Usnea is sometimes called "apothecary's beard", derived from its medicinal uses in wound dressings, general health tonics and whooping-cough remedies. The medieval doctrine of signatures held that ailments could be treated by natural substances that looked a bit like either the affected body part or the disease itself, so *Usnea* was also thought useful for scalp problems, dandruff and leprosy. Lungwort (*Lobaria pulmonaria*) looks a little like human lung tissue, so was deemed good for respiratory conditions. Dove-grey *Parmelia* (crottle) is more sinisterly known as "skull lichen", because if it was found growing on a skull, it was believed effective in treating epilepsy. Dog lichen (*Peltigera canina*), sporting fang-like underside growths, was administered to dog bites and, even more optimistically, to cure rabies.

Lichens have been used in folk traditions for their antibiotic, antifungal, anti-inflammatory, antiviral, antitumour and antioxidant properties, either ingested or administered topically. Some research has been conducted on their possible pharmaceutical uses and is showing promising results. This work is in its infancy, but the scientific world is waking up to centuries-old customs. The therapeutic potential of lichen may yet be useful in new treatments, but while some thrive in extreme conditions, others are sensitive to pollution and may be lost as our climate changes.

Lichens form a vital part of the ecosystem, feeding large and small creatures alike. From snails, butterflies and termites to birds, squirrels, camels and even crabs, the animal kingdom needs them. Reindeer survive long winters digging through deep snow to reach *Cladonia rangiferina* (reindeer lichen). Each animal munches 3 to 5kg per day, just about sustaining its life with this meagre, scaly meal.

Lichens are extremely long-lived; some specimens are hundreds of years old. They can dry out and rehydrate with the seasons and sustain themselves through lean periods. What they cannot do is withstand the extremes of climate change. Recent wildfires, the fires of 2021 in particular, have become a serious threat – for example, to the native species of California (ironically the only US state with an official State Lichen), the lace lichen, *Ramalina menziesii*. The canaries in this particular coal mine are quietly slipping into unconsciousness and not enough of us seem to be noticing.

Right Bighorn cup lichen (*Cladonia cornuta*) from G.F. Hoffmann *Descriptio et adumbratio plantarum e classe cryptogamica Linnaei quae lichenes dicuntur*, 1790–1801.

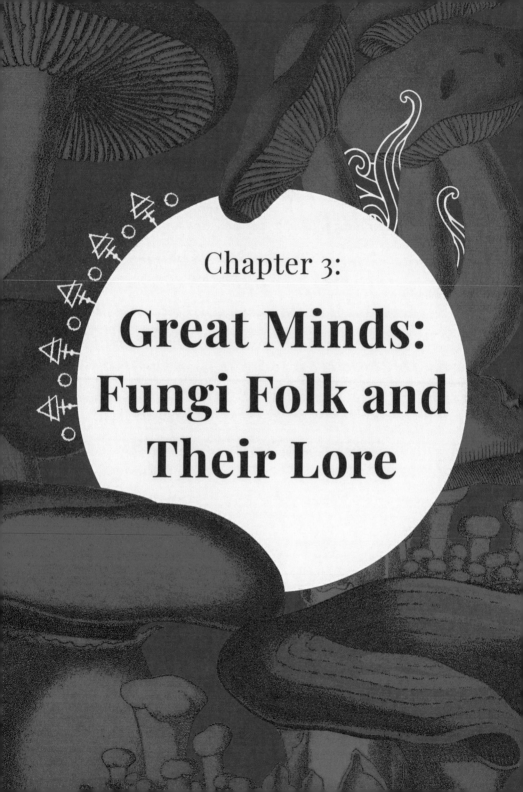

Chapter 3:

Great Minds: Fungi Folk and Their Lore

From the playwright Euripides and biologist Theophrastus to author and illustrator Beatrix Potter and musician John Cage, humans have treated the fungal kingdom with wide-eyed, often obsessive, fascination. We will never know the names of many great mycologists, the true amateurs who learned about mushrooms through forage, examination, trial and error. Even some of the most famous figures remain shrouded in a similar mystery to their chosen subject.

Thousands of years before the word "mycology" had even been coined, Pliny the Elder (23–79 CE) wrote about mushrooms in his *Natural History* with suspicion, while admitting that some, including truffles, were very good to eat.

Italian botanist Pier Antonio Micheli (1679–1737) described around 900 fungi and lichens for the first time in *Nova plantarum genera*. Coming from a poor background, Micheli was not always well-regarded by his "better-educated" contemporaries. Experimenting with spores on slices of melon, he discovered fungi do not spontaneously erupt from the ground, but reproduce via spores. His notion was generally disregarded by fellow scholars.

Like so many men of the cloth, the Reverend Miles Joseph Berkeley (1803–89) became a serious amateur botanist, but he also ended up naming a new discipline. Living and researching on a meagre clergyman's stipend, Berkeley coined the word "mycology" while describing around 6,000 species of fungi for Sir William Hooker's *The British Flora* (1830). He was particularly interested in fungi that attack vegetable crops, including the potato blight, *Phytophthora infestans*, now considered to be an oomycete rather than a fungus, and pathogenic fungi that colonize as rusts and mildews. Berkeley created a huge herbarium of over 9,000 specimens, which now lives at the Royal Botanic Gardens, Kew.

Not all mycologists were clergymen, of course. Anna Maria Hussey (1805–53) didn't take too kindly to playing the role of a regular vicar's wife, and was happier researching and illustrating fungi than ministering to parishioners. Hussey both contributed specimens to her friend Reverend Berkeley's herbarium and uncredited illustrations to other people's books while compiling her own magnum opus, *Illustrations of British Mycology, Containing Figures and Descriptions of the Funguses of Interest and Novelty Indigenous*. The first volume did not merely contain 90 exquisite colour plates – it was also a compendium of Hussey's thoughts and observations regarding the collection and study of fungi. She even included advice on how to hunt for specimens, which still carries a ring of bitter, personally witnessed authenticity: "A basket is in the first place needful, and if the student should leave home without one, a profusion of lovely and rare objects will be certain to strew his path." A second volume was truncated by Hussey's death but, published posthumously in 1855, still contains 50 magnificent plates.

Hussey corresponded with several leading mycologists, including Mordecai Cubitt Cooke (1825–1914), who led a life as interesting as his name. Like many mycologists before him, Cooke's education was scanty, but he received help from an uncle who instructed him in botany, mathematics and languages. At different points, he worked as a lawyer's clerk, school teacher, journalist, Kew mycologist and curator of the museum of the East India Company. Always trying to keep up with his bills, Cooke published constantly – including, as "Uncle

Opposite Detailed drawings of fringed sawgill mushroom (*Lentinus crinitus*) by J. Sowerby, from a paper read by Reverend M.J. Berkeley to the Linnean Society, 18 February 1845.

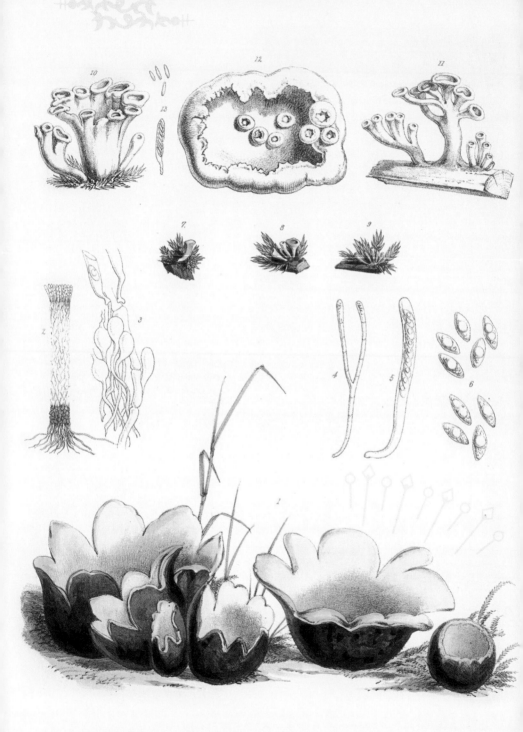

Matt", a general-knowledge series for young readers – but his personal passion was mycology. *The Seven Sisters of Sleep* (1860), which some consider the first of the "psychedelic fungus" books, was followed by *A Plain and Easy Account of British Fungi* (1862) and *Edible and Poisonous Mushrooms* (1894). *Illustrations of British Fungi* (*Hymenomycetes*), published in 1881, was illustrated by Reverend Berkeley's daughter, Ruth Ellen Berkeley, a respected mycologist in her own right.

While botany was considered an acceptable pastime for women, mycology was thought most unfeminine, especially in nineteenth-century America. Mary Banning's (1822–1903) Baltimore neighbours assumed the "toadstool lady" they saw boarding trolley buses, her basket laden with weird and often slimy fungi, was crazy. In reality, she was on a mission to document every species of fungus in her state. After her father died, Banning had to earn money as a schoolteacher to take care of her mother and older sister, but she never lost her soul. Over nearly 20 years, she painstakingly worked on *The Fungi of Maryland* in what little spare time she had.

People didn't trust mushrooms and they certainly didn't trust a woman with them. Even within her chosen discipline, Banning was made to feel inferior. She had just one mycological "friend", the New York State Botanist Charles Horton Peck (1833–1917), with whom she corresponded but never met. In 1888, she sent her entire manuscript – 175 hand-painted watercolours and corresponding descriptions, including 23 new species – to Peck, hoping for peer review and, eventually, publication. For many years, she wrote repeatedly, asking what he thought of the

Opposite Lichen and fungi from a paper by Reverend M.J. Berkeley published in the *Transactions of the Linnean Society*, 1845 including cedar cup fungus (*Geopora sumneriana*).

manuscript; it is unclear whether he ever replied.

Banning died in a cheap boarding house in 1903, her work still lodged in the Herbarium of New York State Museum. In the 1910s, mycologist Howard Kelly consulted the work, then it languished until the 1980s when curator John Haines found the manuscript in a drawer. Haines was staggered by its beauty and scholarship and, at last, Banning's work was recognized. Her book, however, remains unpublished.

Most of mycology's greatest exponents studied the biology of fungi. Valentina and Robert Gordon Wasson were interested in a different aspect of the kingdom: its relationship with humans.

Dr Robert Gordon Wasson (1898–1986) was born in Great Falls, Montana. He spent many of his teenage years in Europe, following a series of eclectic paths, studying with archaeologists, economists and linguists, becoming a journalist and eventually a banker, but his real passion began when he met and married Russian physician Valentina "Tina" Pavlovna Guercken (1901–1958) in 1926. While walking in the Catskills on their honeymoon, Gordon was astonished when Tina fell to her knees and pounced on a small colony of mushrooms. To his horror, she brought a skirtful back to their lodge, cleaned, cooked and ate them. He refused to partake of this deeply suspicious feast and famously admitted he fully expected to wake the next morning a widower. Instead, Tina, who came from the Russian tradition of mushroom foraging, converted her husband to the concept of fungi as something to enjoy and study.

The couple pooled their wide range of experiences and skills to build a new anthropological picture that included archaeology, taxonomy, comparative religion, folklore, art and literature. To this, they added their own personal experiences: Tina's childhood within a culture that lived for its love of mushrooms;

Gordon's polar-opposite memories of having been brought up in the "mycophobic" society of America.

In 1952, encouraged by the poet Robert Graves, the couple began focusing on the mushroom rites of the Mazatec people of Oaxaca, Mexico. María Sabina, a *curandera* (native healer), in the tiny village of Huautla de Jiménez, introduced them to a *Psilocybe* mushroom, and its uses in the traditional medicine and spiritual rituals of her people. In 1955, Gordon Wasson and a friend became the first outsiders to participate in the secretive midnight rites of the cult of the sacred mushroom. A day or so later, Tina repeated the experience.

The "secret" was not to last much longer. Gordon Wasson wrote about his experiences in a famous *Life* magazine article, "Seeking the Magic Mushroom", and became a household name almost overnight. Six days after *Life* magazine hit the newsstands, Tina Wasson's version of the story, "I Ate the Sacred Mushrooms", was published in *This Week*, becoming one of the first times *Psilocybe* was suggested as useful for therapeutic purposes. While 1950s America was perfectly happy, even titillated, to see a man initiated into a magic-mushroom ceremony by a "native woman", it was not ready for a nice American "housewife" to join in. Although Valentina Wasson was the scientist – and lead author on the couple's jointly written, two-volume masterpiece, *Mushrooms, Russia and History* – even today, it is generally Gordon who receives the accolades.

A groundbreaking and comprehensive volume, *Mushrooms, Russia and History* (1957) explored the folklore, literature, art and culture surrounding fungi, spawning a new discipline, "ethnomycology". The Wassons also pointed out something that no one had admitted before: that the world was roughly divided into "mycophiles" – the mushroom-loving

Russians, Eastern Europeans and Catalans – and "mycophobes", the fungus sceptics, including Greek, Celtic, Scandinavian and Anglo-Saxon cultures.

Valentina Wasson died from cancer in 1958, but Gordon was determined to continue their work. He retired from the bank in 1963 to continue his research, publishing, with Dr Roger Heim, *Les Champignons Hallucinogènes du Mexique* in 1967, and, with Dr Wendy Doniger O'Flaherty, *Soma: Divine Mushroom of Immortality* in 1969. Virtually all the fungi folklore known today has been winkled out by Gordon and Tina Wasson, or someone following directly in their ethnomycological footsteps.

María Sabina (1894–1985) had a less happy fate. Her secluded village became an unwilling staging post on the 1960s hippy trail as first celebrities and then tourists plundered her "children" (mushrooms), which she had always considered to be agents to cure the sick, not fuel a hedonistic lifestyle. Horrified at the disrespect shown to their culture by raving tourists who would ingest *Psilocybe* then rampage naked through the village, Sabina's neighbours blamed her for the Western "pollution" and burned down her house. The Mexican police treated her as a drug dealer. Only after the army closed the village to all but locals was the community able to begin to heal.

Gordon Wasson later expressed remorse for his role in revealing the ritual, but it was too late for María Sabina. She died penniless in 1985.

There is some hope. Numerous studies into the therapeutic benefits of psilocybin are beginning to show benefits in the reduction of anxiety and depression in long-term cancer patients. María Sabina may yet prove to have been right all along.

Opposite *Amanita strobiliformis* (warted amanita), illustrated by Ruth Ellen Berkeley from Mordecai Cubitt Cooke *Illustrations of British Fungi*, 1881.

AGARICUS *(AMANITA)* STROBILIFORMIS *Fries.*

on the ground. (*near King's Lynn.*)

Penicillin

Penicillium

Every schoolchild learns of the happy day in 1928 when Alexander Fleming returned from holiday, noticed fungus growing over a petri dish of bacterial culture he'd forgotten about and, abracadabra! Penicillin was invented.

This Hollywood version cuts out years of painful slog, for the magical reveal, missing a richer story in the process.

Penicillium has been with us since we first evolved. Millennia before anyone would utter the word "antibiotic", ancient Egyptian medics placed poultices of mouldy bread on wounds after noticing the curative effects. Chinese, Grecian, Roman and Serbian physicians were finding similar results.

Herbalist John Parkinson's *Theatrum Botanicum* (*The Theatre of Plants*, 1640), reminded bakers to "mould-up" their bread so it would rise more easily. Mouldy bread was a folk remedy across Europe and America, though in rural Quebec, folk preferred mouldy jam. Scholar Frank Dugan notes that in the 1950s, the *Sunday Times* ran a column on folk cures, consisting of all manner of mouldy items, including straw, cheese, apples and leather. Mould didn't always come from food, and was considered especially potent if gathered from churchyard graves.

In the scientific world, it was British microbiologist John Burdon-Sanderson (1828–1905) who first noticed bacteria did not grow near *Penicillium*. Scientists across Europe studied the phenomenon, including Joseph Lister and Louis Pasteur. In 1897, French military physician Ernest Duchesne presented his theory that an antagonism between bacteria and moulds could be used for therapeutic purposes. He came up with the idea by observing Arab stable boys treating saddle sores with mould from saddle leather. Unfortunately, Duchesne used a weak member of *Penicillium*, *P. glaucum*, and although he had the right idea, the moment was not his.

Alexander Fleming's eureka moment was the result of years of searching. His breakthrough mould was *P. notatum* (now known as *P. chrysogenum*). In 1939 Ernst Chain, working with Howard Florey and Norman Heatley, isolated penicillin from Fleming's "mould juice", but couldn't reproduce enough to save their first patient. In 1943, American laboratory assistant Mary Hunt noticed a golden mould, *P. chrysogenum*, growing on a cantaloupe melon, paving the way for mass production of the life-saving drug.

Fleming, Florey and Chain received the Nobel Prize. Hunt received a nickname: Mouldy Mary.

Opposite Colonies of *Penicillium rubens* growing on a culture plate. Illustration by Katie Scott, from *Fungarium*, 2019.

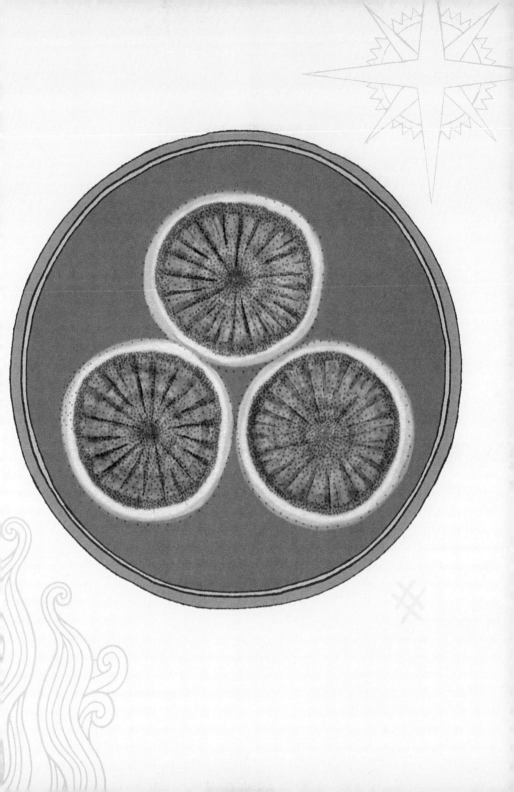

Chapter 4:
Fairy Rings

Fairy rings – ever-increasing circles of mushrooms that suddenly appear in a field or lawn – are probably the most pervading symbol of mushrooms in folklore. Virtually every country has its own interpretation (or interpretations) of what they mean and how they should be approached, if at all.

Whatever they are known as, most traditions agree that fairy rings are magical but, unlike many ancient beliefs that they were somehow connected to thunderstorms, more recent superstitions suggested they were the work of supernatural beings or evil intentions.

In the Netherlands, the circles grow from the splashes left when the devil churns his milk. In Austrian folklore, they mark the place where a dragon's tail scorched the earth – after such a battering, nothing but toadstools may grow there for seven years. In the Philippines, the rings are made by tiny spirits, while in Hawaii they are known as Menehune rings, after the legendary little people that secretly build things at night on the islands. A common theory across Europe holds that such circles are the result of witchcraft. In Germany, *hexenringe* ("witch rings") are the equivalent of the French *ronds de sorcières* and the English (Sussex) *hag-tracks* – evidence that witches have been dancing, especially on Walpurgis Night (May Eve) or All Hallows' Eve (Halloween).

Whoever – or whatever – is responsible, the most popular explanation for fairy rings across the world is dancing. Native Americans suggest the rings are caused by dancing bison. In Scandinavia, *elfdans* are caused by dancing elves, though not always as daintily as we may assume. In 1555, the Swedish writer Olaus Magnus suggested the elves burn the ground with their feet. He explains in *Historia de Gentibus Septentrionalibus* (*A Description of the Northern Peoples*) that the soil has to be refreshed by the mysterious trickster spirit Puck. Magnus's book was hugely influential throughout Europe, translated into several languages and even ended up at the English court, but many of his ideas had already

reached British shores. In twelfth-century Middle English, *elferingewort* roughly translates as "a ring of daisies caused by elves' dancing".

The dancing fairies of Britain and Ireland are not always the cute little creatures beloved of Victorian illustrators. They cavort on moonlit nights, their tracks only becoming visible the next morning. The mushrooms around the outside are small seats for the weary or for spectators, though absolutely not of the human variety. Some stories talk of humans joining in the dance, with mixed results. Most traditions think even approaching such a liminal place as a fairy ring is an act of rank foolhardiness – and not just because in France they are guarded by bug-eyed toads. Consequences of stepping inside the circle range from merely dying young, going lame, losing an eye, being struck completely blind or getting hanged if one is a thief (in Somerset, the ring is known as a galley or gallows trap) to becoming permanently invisible to other humans and being held captive by elves. This might mean disappearing into perpetual slavery in the magical kingdom below the ground or being forced to dance to exhaustion, madness and even death through fairy malice.

The fair folk will often try to lure an unsuspecting mortal into their dance, appearing both comely and

Opposite Fairy ring mushrooms (*Marasmius oreades*) by Kew mycologist Elsie M. Wakefield, Kew Collection, c.1915–45.

E.M.Wakefield

cheery, dancing to a harp and beckoning the traveller, or shepherd, or journeyman, or lover to join them in their deadly whirl. They might offer temptations in the form of food, riches or the attentions of a stunningly beautiful fairy; these must be avoided at all costs.

Once caught in the ring, the victim must often dance for a year and a day before they can be rescued. Welsh lore suggests throwing marjoram and thyme inside the circle to confuse the fairies, or touching the victim with something made from iron, which supernatural beings cannot bear. Staffs made from the holy rowan tree (*Sorbus*) also might work. Pulling the victim from the ring was an involved business; one

farmer from Llangollen tied a rope around his waist and got four burly men to hold it as he entered the circle to rescue his bewitched daughter.

In some traditions, even after they are freed, victims have no memory of their experience – or discover that in the brief time they spent in the magical realm, eons of human time have elapsed. In one tale from Carmarthenshire, a man returns to his own time only to crumble to dust. Another collapses after he tastes human food. Fairy music from the night's exertions may haunt a victim forever, or they might waste away in the human world after experiencing the "other place".

Above Nineteenth-century illustration of a fairy
ring, by children's book illustrator Kate Greenaway.

However distressing fairy rings might be, folklore is pretty much unanimous that it is pointless to try to destroy one. It will only grow back – and bigger. In Irish legend, to damage a sacred circle is to anger the gentle folk, who will curse the perpetrator even if done by accident, perhaps by a farmer's plough. In Worcestershire, kicking a mushroom invites seven years of bad luck. Much better to work with the magic and glean what one can from the fair folk.

Anyone contenting themselves to run around the outside of the mushroom ring nine times in the direction of the sun (to run "widdershins", or anticlockwise, is to invite trouble) will be able to hear the fairies who live underground. The custom is most popular in Northumberland but can be found across the British Isles. Some traditions suggest the laps are best done by the full moon or on May Eve or All Hallows' Eve, when the veil between worlds is at its thinnest. They should not overdo things, though – to complete a tenth lap brings bad luck. If someone really must step inside the circle, they should place their foot on that of a friend standing outside. They might also try wearing their hat backwards. Our "good neighbours" may be cunning, but they are easily confused by displays of human eccentricity.

Treated with care, fairy rings can be beneficial. They can mark the site of buried treasure, though would-be adventurers usually need supernatural help to uncover it. Young girls, if they can acquire some without entering the ring, might splash their faces with dew from within the circle to improve their skin. But if they want to use it to bewitch a would-be lover, they should be careful not to get any on their faces during the spell – or it will maim their sweetheart's face instead.

Although some traditions hold that animals refuse to graze within a mushroom circle (and if they do, their milk curdles), in Wales, mountain sheep that eat such grass thrive. In other places, crops are said to be best where a ring once stood. In Ireland, mushrooms are thought to be shy, and hide away once they have been viewed. Spotting a new fairy ring can, for some, bring fortune and fertility but, in general, positive traditions around fairy rings are much rarer than those promising woe.

Dancing is not the only explanation for mushroom circles. In the Isle of Man, they mark the site of an underground fairy village. Shooting stars, lightning and meteorites all account for the sudden appearance of such circles, while alternative explanations involve the urine, spit or semen of elves and, on occasion, stallions. Indeed, stallion semen was a sure-fire way of encouraging mushrooms if you really wanted some to grow.

In Scotland, it's been said that mushrooms are fairy dinner tables; in Wales, they are fairy parasols. The story is echoed in Central America, where children are sometimes told that mushrooms are umbrellas carried by woodland spirits, who leave them behind at dawn when they retreat to their own world. A few legends talk about what may be found inside a fairy ring, such as the Devonian story that a black hen and chickens appear at twilight in a certain fairy ring on Dartmoor.

Scientific explanations

Searches for a scientific explanation of fairy rings began as far back as a paper given to the Royal Society in 1675, when it was suggested – much as the ancients had done – that the rings were "electrical" in origin, the result of a lightning strike.

In 1807, English chemist William Hyde Wollaston suggested they had "chemical" origins. Later came a "mineral theory", followed by the somewhat less savoury "excretory theory". None really seemed to fit the bill, though the agricultural research institution Rothamsted put forward a "nitrogen theory", whereby the fungi were somehow managing to extract nitrogen from the soil. It felt as though the scientists were getting warmer.

Only relatively recently have mycologists been able to prove exactly what is going on with a fairy ring. The bit we can see is only a small part of a much larger organism, a thread-like, underground mass (the mycelium) that is present throughout its life and suddenly bursts forth with fungi at certain times of year. This is often after heavy rain, which may explain why so many people thought the "magic circles" had something to do with thunderstorms. Greener, faster-growing grass appears around the outside of the ring thanks to nutrients released from the fungus, while the area underneath the inside of the ring can become choked with mycelium, cutting off the water supply and making the grass dry. When farmers noticed animals preferred the pasture outside the ring, they may have assumed it was because the inside was evil – rather than the simple fact that the outside just tasted better.

Over 100 species can be ring-forming, including some highly toxic toadstools, but the most common is *Marasmius oreades*, or fairy ring champignon, sometimes known as the "resurrection mushroom" due to its ability to regenerate after having dried out. The rings can reach gigantic proportions – one entirely encircles Stonehenge and is roughly a thousand years old. Such rings may account for legends surrounding specific ancient sites. Dartmoor has several stone circle sites that were believed to be home to dancing pixies. Some so-named fairy

Right *Ring of Fairies Dancing in the Moonlight,* watercolour by George Cruikshank.

rings may have nothing to do with mushrooms. For example, the strange-looking Fairy Ring, aka Table des Pions (Pawn's Table) in Guernsey, is a circular earthen dugout locally associated with evil fairies. It was actually built as a meeting place for island officials, who would sit with their feet in the ditch to eat their lunch in between inspecting roads and coastal defences, up to 1837 when the last official inspection took place.

Fairy rings may be "free" or "tethered", depending on their food source. Free rings derive their nutrients from organic matter taken directly from the soil, and are often found in open grassland. *Marasmius oreades* is a typical example, growing centrifugally (outwards in a circular habit) and creating a new ring of fruiting bodies each year. Each grows a little wider, while the centre dies. Tethered rings are usually found in woodland and depend on a specific food supply. Far from being dangerous to the tree, it is believed that this mycorrhizal relationship is mutualistic, actually helping the tree to absorb moisture and nutrients.

Fairytale mushrooms

Humanity has always had an uneasy relationship with mushrooms, not least because it is so easy to pick the wrong one and die an unpleasant death. Legends arose in many cultures as warnings to curious children against "testing" interesting growths they found in the woods.

A common idea that warty, baleful toads sit on certain fungi waiting to trap flies, for example – in German, *krötenstuhl* literally means "toad stool" – may have been a way to put them off the idea. The Ozark people of North America warn that to pick fungi at any other time than a full moon will render the food unpleasant or even poisonous. This duality, drawing on the good/evil aspect of mushrooms, made for ripe pickings for fairy tales, especially in Eastern Europe.

The War of the Mushrooms is a cautionary/comic Russian folk rhyme where King Bolete (sometimes King Pine from *Boletus pinophilus* or pinewood king bolete) orders his subjects to muster for battle against the beetles, but most find excuses not to fight. The fly agarics (*Amanita muscaria*) are far too noble; the morels (*Morchella*) claim old age. Honey mushrooms (*Armillaria*) and ink caps (*Coprinopsis atramentaria*) plead their skinny legs, while other species demand special treatment as poor old ladies, nuns, peasants, professional layabouts or court officials with important jobs. In the end, the mushroom army is chiefly formed from the courageous peppery milkcap (*Lactifluus piperatus*). Beautiful nineteenth-century illustrations survive from the original tsar-era versions; in Revolutionary Russia, the tale was reinvented as propaganda. Igor Stravinsky set the story to music in 1904, then put it away in a drawer. "How the Mushrooms Went to War" only received its debut in 1998.

The Russian version of the *Cinderella* story has "Old Man Mushroom" playing the role of the Fairy Godmother, and not just to Nastenka, the heroine. Her Prince Charming equivalent, Ivan, is an egotistical neighbour (similar to Gaston in the Walt Disney version of *Beauty and the Beast*). Old Man Mushroom teaches the cocky young man a lesson in humility by turning his head into that of a bear. In order to regain his human head – and the love of Nastenka – Ivan must go out into the world and do good deeds. In a North American version of *Cinderella*, "Ashpet" borrows some fire from an old woman (the Fairy Godmother figure) and carries it home in a bracket fungus. An Italian folk tale sees the other side of a witch's wrath when she disguises herself as a mushroom to catch cabbage thieves in her garden.

Andrew Lang was a Scottish folk-tale and fairy-tale collector working at the end of the nineteenth century. His famous, multi-coloured Fairy Books series (actually translated and retold by his wife Leonora Blanche Alleyne) were supplemented by his own original stories, including *Princess Nobody*, which uses many of the classic fairy-tale tropes. The ugly but kind-hearted Prince Comical must search for a lost princess without a name in the dangers of Mushroom Land. He is warned by a black beetle he must never sleep under a toadstool, a test that, thanks to the insect's warning, he passes. He discovers the princess's name, but after having been made handsome as a reward by the Queen of the Mushrooms, he fails the next test by calling the Princess by name and has to act fast to save the day.

Raymond Briggs's 1977 *Fungus the Bogeyman* might be considered a modern-day fairy tale, that of an ordinary bugaboo in a topsy-turvy world where all things "nasty" – including mould – are nice.

Above Fairy ring mushrooms (*Marasmius oreades*) by Anna Maria Hussey from *Illustrations of British Mycology*, 1847–55.

How mushrooms were created: a folk tale

Jesus and Saint Peter were wandering the countryside when they came to a poor village. From one of the smallest cottages, they heard the inhabitants celebrating a wedding.

Not knowing the identities of the strangers, the couple invited Jesus and Peter to join them and share their food. Jesus warned Peter that these people were poor and that he should not eat anything but a little bread and salt.

The pair joined the party and ate some bread and salt, but Peter could not help sneaking some delicious little koldce (cakes) into his pouch. Soon the two departed, and after a while, Jesus noticed that his friend kept holding back and nibbling something. Every time he asked what he was eating, Peter spat out his mouthful and muttered "nothing." This continued until there was no stolen cake left.

Jesus told his friend to go back and pick up all the pieces he'd spat out. Peter retraced his steps. But instead of morsels of cake, he found strange growths. Jesus told him the weird shapes were mushrooms that had grown from the peasants' food, which Peter had stolen and then thrown away. Ashamed, Peter begged for forgiveness, which Jesus granted. They stopped at the house of a poor woman and asked her to cook the tasty mushrooms. Jesus gave them to the poor, from whose food they had sprung.

He made sure that mushrooms are always plentiful, but because Peter remained hungry after trying to eat the cakes, fungi will never be a filling meal.

Versions of this story are told across Eastern Europe; a rendering of the tale above was first recorded by Czech folklorist Božena Němcová (1817–62). In another version, Peter and Jesus are walking through a rye field and Peter sneaks grains into his mouth against Jesus's express instructions. Sometimes the holy pair are seen begging for bread, and the fungi spring from crumbs they drop. White breadcrumbs become mushrooms; brown crumbs turn into toadstools. Occasionally, no food is involved at all; in Sicily, Hungary and Lithuania, the mushrooms magically appear out of Peter's spittle. Such stories may originate from earlier tales of the Lithuanian god Velnias, whose fingers reach out from the underworld to feed the poor.

Opposite Fairy inkcap (*Coprinellus disseminatus*) from James Sowerby *Coloured Figures of English Fungi or Mushrooms*, 1795–1815.

Sept. 1. 1798. Published by J. Sowerby London.

Agaricus striatus

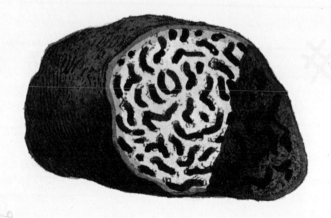

Tuber moschatum

Agents of thunder

In almost every culture, mushrooms have, at some point, been associated with thunder, lightning and rainstorms. The mythology is so ubiquitous that some have begun to wonder if there might be something in the tradition.

These beliefs go way back. Roman writer Pliny the Elder recommended seeking out truffles immediately after storms, "thunder contributing more particularly to their development". The god Jupiter (Zeus in Greek mythology) was in charge of the weather; he was who the Romans prayed to for a good mushroom harvest. Pliny even pointed out that the best truffles come from Olympia, Zeus's sacred realm, where he could throw as many thunderbolts as he chose. The Roman poet Juvenal ached for a spring thunderstorm to bring him fungal riches.

Valentina Wasson conducted much investigation into lightning and mushrooms, telling us that, in Russian, a warm, soft rain is *gribnoy dozhd*: "mushroom rain". She talks of a legend in the Zarafshan valley of Central Asia where a goddess known as "Grandmother" shakes her baggy trousers, sending a swarm of lice to the ground. After thunderstorms, the lice turn into mushrooms, an appealing thought in mycophilic countries. The Himalayan explorer Charles Evans (1918–95) noticed that his Sherpa guides, who came from Tibet, also believed that thunderstorms make mushrooms appear.

Opposite Black truffles (*Tuber melanosporum*) from James Sowerby *Coloured Figures of English Fungi or Mushrooms*, 1795–1815.

A Bedouin tradition holds that a severe season of autumnal electric storms will bring a large crop of *Terfezia* the following spring. Not all years will be good, but if the storms are heavy in October, it is worth packing a tent and moving to sites known for generations as yielding a good crop. Women and children are deemed best at detecting the faint shadows cast by the swelling underground fruiting bodies in the long first rays of the dawn or the dwindling twilight. The mushrooms are soaked in salted water and fried in butter or roasted in hot ashes and dipped in salt. In a good year, there will be enough to sun-dry for the hot summer months.

Across the Philippines, people also make their way to open meadows after spring storms to gather "thunder mushrooms". Wasson thinks the belief may have spread from Malaysia, India or even China, all of which associate mushrooms with thunderstorms. She cites the Chinese "thunder-aroused mushroom", which she suspects may be the morel (*Morchella*), "thunder-peal mushroom", and "thunder mushroom", neither of which she hazards a guess as to species. She does note, however, that for the Māori people of New Zealand, the words for thunder and mushroom are the same: *whatitiri*, which also happens to be the name of a mythical ancestor-goddess, the "thunderer".

The relationship between lightning, thunder and fungi crosses all continents, including the First Nations of North America, and the Mesoamerican mushroom cults of Mexico and Guatemala. The Nlaka'pamux First Nations people of British Columbia call shaggy ink cap mushrooms (*Coprinus comatus*) "the thunderstorm head".

With such widespread belief — spanning cultures that, for many centuries, could not have known each other — perhaps there is something in the mythology? Certainly, most fungi need moisture to survive. They need a lot of moisture to flourish. Since thunderstorms usually come with plenty of rain, it seems reasonable that extra rain equals extra mushrooms. The vast amounts of rain in certain parts of the world in recent years means some places have seen record crops of fungi. But storms?

Raijin is the Japanese trickster god of storms and general chaos. He may look demonic, but he is celebrated as the drummer-up of thunder, bringer of rain and friend to the farmer. However, he sometimes either slacks in his duties or, more often, finds himself in prison for various misdeeds. This means drought, and a dip in the mushroom harvest. There are worse things than a lazy god. In recent years, a nematode has been destroying the red pine forests that grow some of Japan's favourite mushrooms.

Japanese scientists are leading the research into possible connections between electrical storms and the much-anticipated mushroom harvest. In trying to find ways of growing commercial crops of ectomycorrhizal mushrooms — the sort that normally live in harmony with a tree host in forest conditions — scientists at Iwate University, Morioka, led by Professor Koichi Takaki, revisited old farmers' lore, which welcomes lightning as a way to increase their foraging yield. The research team built their own electric storm machine and zapped Raijin-strength "lightning bolts" around laboratory crops of 10 different species of mushroom. Eight of these saw an increase in growth, the strongest of which was observed in shiitake (*Lentinula edodes*) and nameko (*Pholiota microspora*). At Kyushu University, Fukuoka, Ferzana Islam and Shoji Ohga worked with their own Small Population Lightning Generator. Leaving the laboratory, they trundled the device through a forest, sending 50,000-volt pulses through the ground via an electrode. The hope was to boost harvests of the highly sought-after matsutake (*Tricholoma matsutake*). The experiment enjoyed similar results to those of the Iwate team, and saw even higher yields several weeks later. The weight and size of the mushrooms increased, too.

A genuine direct-hit lightning bolt, at around a billion volts, would completely destroy a fungus, but the weakened charge from a strike further away appears to travel through the soil and stimulate the mycelium into fruiting. Experiments with different strengths of "lightning bolt" have produced double-sized crops, and current speculation suggests that the same urge to reproduce that plants have when faced with adverse conditions (for example, lettuces bolting during droughts) is at work in the fungal kingdom.

More research is needed before farmers can habitually electrocute their mushroom harvest into a bumper crop, not least the invention of a more user-friendly lightning-bolt generator. In the meantime, for the sake of all Japanese mushroom fans, Raijin needs to pull up his *tabi* socks.

Opposite Morels (*Morchella* spp.) from J.V. Krombholz, *Naturgetreue abbildungen und beschreibungen der essbaren, schädlichen und verdächtigen schwämme*, 1831–46.

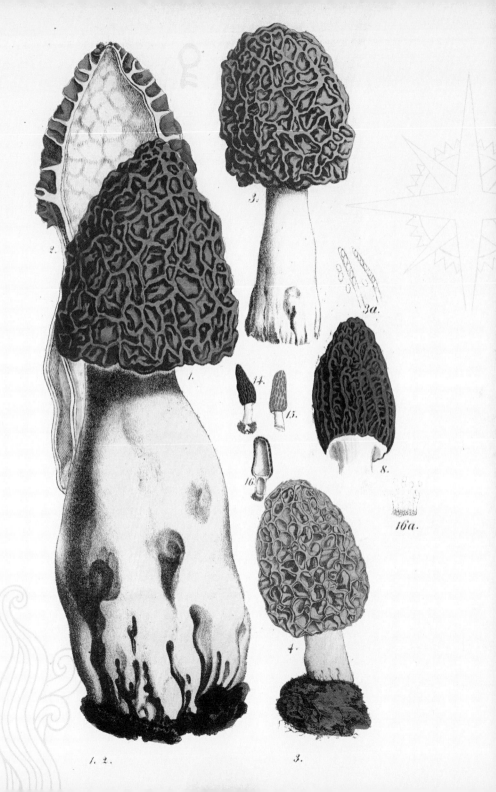

Earthstar
Geastrum triplex

*The extraordinary group of fungi popularly known as
earthstars is not a single genus and has, over the centuries,
been known by many scientific names.*

Regular folk had no such quandaries. They knew exactly where earthstars came from: the heavens.

Found in every continent except Antarctica, *Geastrum triplex* is strange enough to stop a walker in their tracks. It is easy to understand why the Blackfoot people in the Great Plains of America called the specimens that seemed to spring up overnight *ka-ka-too*, or "fallen stars", and believed them to foretell supernatural events. The Cherokee put them in the navels of babies to aid healing from the umbilical cord, while the Tewa people of New Mexico blew spores into the ears of those suffering earache or hearing problems. Even po-faced scientists were not beyond fantasizing upon the romance of sky fall. In 1847, mycologist Charles David Badham wrote, in *The Esculent Funguses of England,* "*Geastrum*, aspiring occasionally to leave this earth, has been found suspended, like Mahomet's coffin, between it and the stars, on the very highest pinnacle of St. Paul's."

The Latin name *triplex* describes the three-layered peridium (outer skin) that forms around what looks like a regular, if slightly brown, puffball. The skin hardens, splits and flattens into a segmented dish shape for the fruiting body to rest on. It is sometimes known as the collared earthstar or the saucered earthstar. It appears mysteriously in a number of habitats, including the detritus of a broadleaf forest, sand dunes and even hedgerows. These curiosities can be large – 10 to 12cm across – and are not generally considered edible. In traditional Chinese medicine, *Geastrum triplex* is used to staunch bleeding, reduce swelling and ease inflammation. When *Geastrum fornicatum*, the arched earthstar, was first recorded in 1671, Dr G. Seger considered it an "anthropomorphus". He illustrated it as a group of humanoids with giant heads and curling, tortured bodies released from imprisonment in the fungus, a bit like the medieval mandrake. In reality, the fruiting body is lifted by its star-shaped former shell, a little like a ballerina on pointe, so the spores can be more easily borne away by the breeze.

There are dozens of species of earthstar, and many bear little or no scientific resemblance to *Geastrum*. The barometer earthstar, *Astraeus hygrometricus*, has long been known as a weather-forecasting aid; it opens up when there is rain and closes to preserve moisture and avoid predators on dry days.

Opposite Fringed earthstar (*Geastrum limbatum*) by Anna Maria Hussey from *Illustrations of British Mycology*, 1847–55.

Geaster limbatus, Fries.

Fungi and witches

Strange, mystical, and with a sinister habit of appearing and disappearing as if by magic, mushrooms enjoy a long association with witchcraft.

Witches were traditionally held responsible for "evil" fungi. Moulds, blights and rusts that attacked crops were clearly evidence of a curse, while obscene-shaped mushrooms that sprang up in the gardens of good Christians had probably, it was said, been conjured by a slighted sorceress. In one well-known 1926 court case in Melun, France, two men and eight women of the Notre-Dame des Pleurs cult were accused of attacking an abbot they believed possessed by the devil. Stripping him naked and beating him with knotted ropes, they claimed he had sent evil birds to defecate on the garden of the sect's founder, Mme Mesmin. The fungus that sprang from the birds' droppings looked and smelled so offensive that anyone who breathed it fell sick with a "shameful disease". It was whispered that Mme Mesmin had wanted rid of the clergyman as she feared he was a more powerful sorcerer than herself.

Fungus is an obvious ingredient for potions. Fly agaric (*Amanita muscaria*), often claimed as a vital part of flying ointment, is known as *hexenpils* ("witch

mushroom") in Austria. Puffballs were denounced by the Inquisition as ingredients for dark concoctions across the Basque region of France. The innocently named petticoat mottlegill (*Panaeolus papilionaceus*) is a dung-loving, inedible mushroom that appears to have been added to Portuguese potions until relatively recently.

The terrifying witch of Russian folklore, Baba Yaga, is heavily associated with mushrooms. They grow around her enchanted house, which lurks in the forest on chicken-footed stilts surrounded by skulls. Naturally, she knows the creatures that live under the toadstools, too. While sometimes cruel and even, on occasion, cannibalistic, Baba Yaga is not always evil. A hero or heroine encountering her is just as likely to be aided as eaten. In one famous tale, Baba Yaga meets a hedgehog in the forest. Both are mushroom hunting. She intends to eat the animal, but eventually spares him, and he turns into a boy with magical powers. Admittedly, other versions say Prince Dmitri was turned into a hedgehog by the witch. That is the way with folklore.

Opposite Illustration for the fairy tale "Vasilisa the Beautiful" depicting Baba Yaga, the witch of Slavic folklore. Ivan Bilibin, 1900.

Common stinkhorn

Phallus impudicus

From the Latin for "shameless penis", the stinkhorn is generally known for two things: its shape and its odour. In the not-so-distant past, it was also known for something else: witchcraft.

There's no mistaking a stinkhorn. It assaults every one of the five senses, and not always in a pleasant way. Charles Darwin's daughter Etty, clad in a special cloak and gloves, armed with a basket and long black stick, famously hunted them down by the putrid stench of death and faeces emitted by the fungus's cap. She sniffed, she poked, she pounced, flicking the offending mushroom into the basket with her stick, never touching it. She would return and burn her haul in secret, out of sight of servants and corruptible children.

The odour, though foul, was not Etty's main problem with the fungus. For her, the mushroom's shape was obscene beyond Christian redemption; she was not the first person to notice what seventeenth-century herbalist John Gerard called "the pricke mushroom". The devil was habitually blamed for the fungus's appearance, remarkable for its similarity to an (albeit not very healthy) erect penis. Common folk names include "Satan's member", "devil's horn" and "devil's stinkpot". The appearance of a stinkhorn, many believed, was the result of a witch's curse, or marked the spot where a sorcerer had urinated or defecated. The "obscene" fungus at the centre of the witchcraft trial in Melun, France, in 1926 (see page 69) was a stinkhorn.

The fungus starts life as a white, ovoid "witch egg" or "devil's egg", about the same size as that of a hen's. The egg "hatches" into a small mushroom, which rises into a rod-like "member", lubricated by a clear, jelly-like substance to ease the erection. The process is remarkably fast; the shaft can grow up to 20cm in two to three hours. The fully grown cap is covered with gleba, a slimy, olive-green, spore-laden mass that smells of rotting/faecal matter and is highly attractive to flies and slugs. It takes little time for insects to consume it and move on. The mushroom's cap is left a white honeycombed shell, before rotting away. It has done its job. The flies, covered in sticky spores, are busy distributing them to everything they touch.

Given the mushroom's shape, it's hardly surprising that it has been considered an aphrodisiac in several cultures. The folklorist Vance Randolph spent years studying the ways of the people living in the Ozark mountains, which cross the US states of Missouri, Arkansas, Oklahoma and Kansas. In 1953, he wrote that young girls stumbling across a stinkhorn considered it a good omen, and how, in the 1870s, they would strip off and dance around

Opposite Stinkhorn, (*Phallus impudicus*) from William Curtis *Flora Londinensis*, 1775–98.

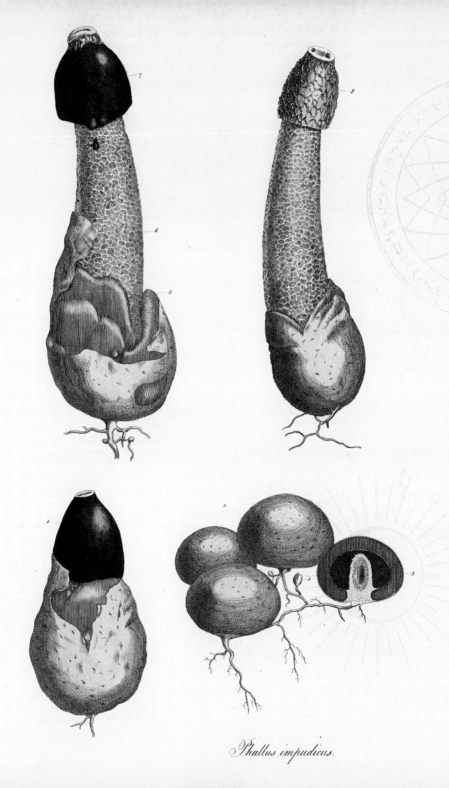

Phallus impudicus.

the fungus. If a virgin touched her vulva with the stinkhorn, she would bag the man of her dreams. Robert Rogers tells us that in Borneo, the stinkhorn is thought to be the spirit-penis of a dead hero. Not everyone was quite so graphic. The slightly more pure-minded Germans thought of it as a finger, perhaps that of a dead body, breaking through the soil as a *leichenfinger* (meaning "corpse finger"). Hunters from the Yoruba people of West Africa made a concoction from the fungus that would render them less visible to their enemies. It would certainly help disguise their human smell to animals.

The Igbo people of southern Nigeria consider stinkhorn a mushroom of loveliness, *éró mma*, from the words for "fungus" and "beauty", showing how much it divides opinion. *Mutinus elegans*, or the elegant stinkhorn, is known by some as the "devil's dipstick". Anything so extraordinary looking eventually finds its way into the medicine cabinet, and stinkhorn has been used to treat anything from epilepsy to gout. In Germany, the eggs were fried with cherries to remove uric acid from the body; in India, the fungus treated typhoid. Latvian medicine recommends stinkhorn for gastric ulcers. Traditional Chinese medicine values it for dysentery. Like so many fungi, stinkhorn is of interest to modern science and is being investigated for therapeutic uses in an impressive range of cancers. A rare, short-lived relative, *Māmalu o Wahine* ("woman's mushroom"), that grows on the lava slopes of Hawaii, briefly attracted attention when it was suggested that smelling it gave some female test subjects "instant orgasms" and/or increased their heart rates. Male volunteers apparently found the odour disgusting. Alas, the sample size, test conditions and anonymity of the mushroom used do not lead to anything like reliable conclusions and considerable doubt has been thrown on the experiment. More research is necessary.

The Phallaceae family is wide-ranging, every member of which is curious. A slimmer, pinker, shinier version, *Mutinus caninus*, is known as a dog stinkhorn. *Phallus multicolor* has a bright-orange lace skirt. The netted stinkhorn (*Phallus indusiatus*) comes with a delicate white lace "skirt", or indusium, that reaches down to the ground; it is dried and used in Chinese cuisine. It was also traditionally used as a divination aid in Mexico, and considered sacred in New Guinea.

Opposite Latticed stinkhorn (*Clathrus ruber*) from from J-L Leveille *Iconographie des Champignons de Paulet*, 1855.

Right Stinkhorns *Clathrus ruber* and *Phallus impudicus* from J.V. Krombholz, *Naturgetreue abbildungen und beschreibungen der essbaren, schädlichen und verdächtigen schwämme*, 1831–46.

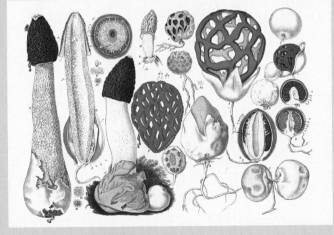

Cup fungi

Given their evocative shapes and bright colours, cup fungi might be assumed to embrace legions of folklore and fairy stories, yet oddly, their legends are few and far between.

Many groups within the fungal kingdom produce cup-shaped sporing bodies, and they are often spectacularly colourful. They are also "spore shooters", meaning they develop reproductive cells called asci on the inside of the cup then, when ripe and the weather conducive, spray them into the atmosphere in a haze of dust. Prone to drying out, they tend to enjoy damp, shady locations.

The bright red scarlet elf cup (*Sarcoscypha austriaca*) translates as "fleshy drinking bowl from Austria", referencing a common tradition that the cup-shaped fungus was used by elves as a goblet. In West Sussex, England, Victorian folklorist Charlotte Latham noted in 1878 that people still regarded the "beautiful red cup moss as the fairies' baths". *Sarcoscypha austriaca*'s downy outer skin is much paler than its deep-red interior, but it remains a striking sight on the woodland floor or clinging to a decaying branch. It particularly likes willow (*Salix*), elm (*Ulmus*), alder (*Alnus*) and hazel (*Corylus*) and feeding mainly on decaying matter in damp, shaded places. It is easily confused with the ruby elf cup (*Sarcoscypha coccinea*) and usually needs a microscope to reveal that the ruby elf cup has straight hairs on its outer surface, while those on the scarlet

elf cup are curled. It is not poisonous, but neither is it particularly nice to eat. Oneida First Nation healers used it to staunch bleeding and placed it inside the navels of newborn babies to soothe the wound.

The equally spectacular emerald elf cup (*Chlorociboria aeruginosa*), is equally inedible and unendowed with folklore, but the wood it infects has been prized for centuries. This stunning turquoise cup fungus looks like exotic shells as it sits on fallen branches, and its mycelium stains the infected wood a striking blue-green known as "green oak". It is popular in marquetry, where small pieces of different-coloured woods are pieced together like mosaic to make intricate designs. The orange peel cup (*Aleuria aurantia*) looks uncannily like a discarded orange skin: similarly found in summer, usually on disturbed ground, as though dropped by a careless tourist. It begins as a cup and, as it matures, the fungus twists and contorts as though the orange has been peeled into strips. Again, it is surprisingly short on folklore, perhaps because although not actively inedible, *Aleuria aurantia* doesn't taste particularly good.

Opposite Various cup fungi including ruby elf cup (*Sarcoscypha coccinea*), figure 8.

Common puffball

Lycoperdon perlatum

The common puffball's scientific name Lycoperdon perlatum *literally translates as "widespread wolf's fart", which is one of its many entertaining folk or common names, and upholds an association with breaking wind that goes back to the ancient Romans.*

The French call puffballs *pet de loup*; another genus , Bovista, is derived from the German for "ox emission". *Ulefyst* is the Frisian "owl's flatus" and *asterputz* the Basque "ass's fart".

"Puffball" is a general term for several species of mushroom that belong to an old form group of fungus called gasteromycetes. These feature reproductive systems that produce spores inside their fruiting body for later dispersal. They can burst open, often from a touch as light as a raindrop, exploding into a cloud of spores. Some split completely; others "puff" short bursts of spores through ostioles ("little doors").

These puffball mushrooms are so common and distinctive that they have acquired dozens of folk names, including "elf fart" and the equally atmospheric "devil's snuff box". Parts of India call the warted puffball (*Bovista pusilla*) *ghundi* or "thunder turd". In Malawi, the chequered puffball (*Calvatia bovista*) is the *ngoma wa nyani*, "drum of the baboon". The umber-brown puffball (*Lycoperdon umbrinum*) is sometimes known in Mexico as *ju'ba'pbich*, or "star-excrement fungus".

The traditional association of fungi with the heavens continues with puffballs, which are sometimes thought to come from the planet Jupiter. They have been worn as magical amulets, filled with gravel as percussion instruments, used in rituals by sorcerers good and evil, prized as tinder or for carrying fire, turned into a paste for wart removal and even puffed into hives to calm bees. Some tribes dry puffballs as incense against evil spirits. In Tibet, puffballs were used to make ink. Another popular association, that of lightning, is seen in the traditions of the Santal people of India and Bangladesh. For them, the *rote putka* ("toad soul plant") is the result of thunderstorms – and possesses a soul.

Depending on one's point of view, certain puffballs may be studded with "gems"– or covered in warts. They may be snow-white, sculpted, tumbling, pear-shaped or even, in the case of *Calvatia cyathiformis*, resemble an old-fashioned loaf of bread. From the *Bovista nigrescens* found at Skara Brae in Orkney and Vindolanda near Hadrian's Wall to the kitchens of trendy modern chefs, true puffballs are our friend. It is important to remember, however, that in their early stages, several highly toxic species can look like puffballs. As with all fungi, approach with caution.

Opposite Giant puffball (*Calvatia gigantea*), from Anna Maria Hussey *Illustrations of British Mycology*, 1847–55. It is edible when young.

Lycoperdon giganteum, *Batsch.*

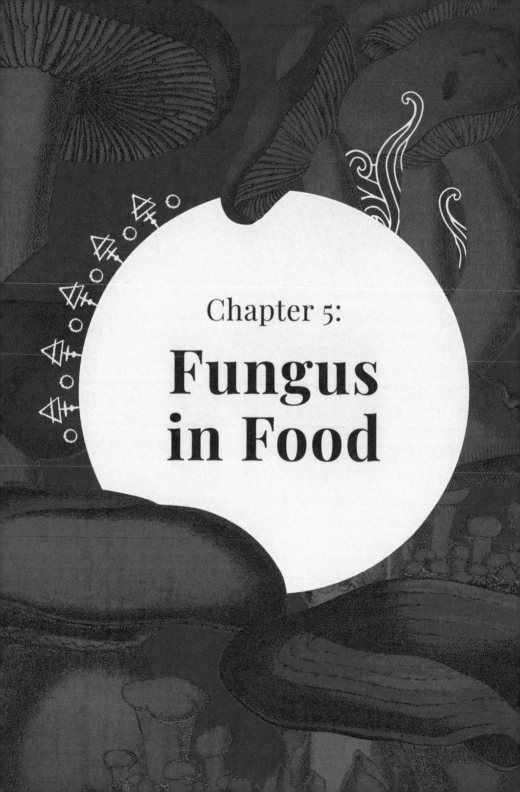

Chapter 5:

Fungus in Food

Humans have eaten mushrooms since prehistory. For millennia, however, we have also been using other forms – such as moulds and yeasts – usually without knowing that they, too, are fungi. Without these cultures, our bread would not rise and our beer would not foam. Bakers, brewers, cheesemakers and other food producers enjoy their own rich folklore.

With the exception of certain highly desired species, fungi meant free food that could be foraged even by people who had nothing else.

Across the world, therefore, mushrooms have usually been considered sustenance for the very poorest. In Tonga, the saying *Mweenzu wa fulwe ulalila bowa* translates as "the guest of the tortoise has mushrooms for supper": don't expect fancy food just because you're a visitor.

Mushroom hunting is a mixture of expertise and luck, so it is hardly surprising that complicated superstitions have sprung up around the practice. In many traditions, if a mushroom is looked at but left for harvesting later – perhaps to grow a little bigger – it will disappear. In Slovenia, you should never mention the mushrooms you're looking for by name, while Californian mining communities traditionally wore one garment inside out to guarantee foraging success.

In Zambia, mushroom gathering is mainly practised by women, who teach their daughters how to tell which may be eaten – and often know more than the scientists. One species, *Termitomyces titanicus*, was only "discovered" in the 1980s, even though it has been eaten and sold in West African markets for centuries. The mushroom can reach over a metre across and is so huge that, in children's tales, animals use it for shelter, including gazelles and green mamba snakes. In reality, it grows on termite mounds and enjoys a symbiotic relationship with the inhabitants. Fireside anecdotes about harvesting the largest examples are not dissimilar to European "anglers' tales".

While it may take the romance out of mushroom collecting, cultivation is a more likely route to success. Miners quarrying rock for the great cathedrals of Paris gradually dug a network of tunnels in the limestone beneath the city's streets. Most famously, they now house the catacombs, filled with the bones of six million Parisians. In the Second World War, the passages concealed members of the Resistance plotting Hitler's downfall, but the nineteenth century saw the reign of the *rosé des prés* ("pink of the fields"), or Paris mushroom. Some suggest that *Agaricus bisporus* was first grown underground by deserters from Napoleon's army; others say it was the brainchild of a farmer named Chambry. The labyrinthine tunnels were nicely damp, kept a constant 12°C (53.6°F) all year round, and the minerals in the rock added a welcome depth of flavour to an otherwise standard mushroom. Alas, the system died in the late nineteenth century, probably with the birth of the Metro subway.

Mouldy food is the ultimate symbol of poverty. In folk tales, the last crust of mouldy bread often represents desperation, the final straw that sends a hero out into the world to find food for the family and/or make their fortune. It is also an indicator of a deserving, humble soul. In the Swedish tale "Lars, my Lad!" a down-on-his-luck duke discovers an abandoned hut in the woods. He opens a wooden chest hoping to find some mouldy bread, but instead discovers a series of ever-diminishing boxes. Inside

Opposite Woodblock print depicting mushroom gathering, by Mizuno Toshikata.

the last casket is a piece of paper that grants him wishes, which he uses wisely to win the obligatory fairy-tale princess.

Mould in the real world is by no means always bad. The fermentation process both preserves food and makes it tastier. *Aspergillus oryzae* breaks down the elements of fermented food products such as soy sauce. Known in Japanese as *kōji* (though the word is also used for other kinds of fungus), *A. oryzae* is first mentioned in a work from second-century BCE China. The *Rites of Zhou* dictates which official is in charge of which aspect of administration, even down to the bureaucrat that had to deal with edible mould.

In Japan, kōji is used to make sake (rice wine) and miso seasoning. Sake is the result of yeast fermentation, but it needs *Aspergillus oryzae* to break down the rice starch into glucose to activate the yeast. Miso is made by fermenting soybeans with salt and *kōji*, imparting a striking flavour and irresistible aroma. Japanese mythology holds that miso was a gift from the gods and brings health, happiness and long life. It was used to lure *binbōgami* – evil *yōkai*, or spirits, in the form of filthy old men – from people's houses into the streets. *Tsukumogami* are household objects that, having reached the age of 100, are gifted with a spirit. One of the most powerful is the *kameosa*, the haunted sake jar, which will never run empty. The connections between sake and the *kuchisake-onna* ("slit-mouthed woman"), Japan's most terrifying modern *yōkai*, are less clear. Perhaps people who claim to have seen the beautiful woman who peels away her surgical mask to reveal a mouth slit from ear to ear have been enjoying the sake a little too much.

Opposite *Huitlacoche*, corn or blister smut of maize, caused by the fungus *Ustilago maydis*. The kernels swell into galls. Drawn from a specimen in Kew's Fungarium by Charlotte Amherst, 2022.

The indescribable taste of umami is mainly associated with Asian cuisine today, but it has been known across the world for centuries. Geo Watkins's salty, spicy mushroom ketchup is the very epitome of umami, yet it was first sold in England in 1830 and is still available today, in bottles hardly changed in design since Victorian times.

Fungal enzymes are sometimes used to replace animal rennet in cheesemaking, but mould is an essential component of blue cheeses, where *Penicillium* is introduced to the culture and allowed to develop deeply pungent blue veins. In Roquefort, France, legend tells a shepherd was in a cave eating cheese for his lunch when he spotted a pretty girl and ran off to find her. By the time he got back, the cheese had gone mouldy. He ate it anyway and discovered fromage nirvana. Italian gorgonzola is sometimes known as "green cheese" for its lighter veins. It is at its best when used with the rich autumn milk produced by cows returning from the mountains for the winter. Both cheeses proudly declare themselves to be the King of Cheese, as does the venerable Stilton of England. Other, non-blue cheeses such as Brie and Camembert are ripened and protected by an edible "skin" of white mould – another incarnation of the ever-adaptable *Penicillium*.

Fungus in cooking is a delicate balance between fortune favouring the brave and caution favouring the living. Mexican diners, for example, have much to thank their forebears for in first testing the *huitlacoche*, known as corn smut "Mexican truffle". This battleship-grey pathogenic fungus attacks corn cobs in large, bulbous growths, theoretically ruining a crop. Its name may derive from the Nahuatl words for "sleep excrescence" or for "raven's excrement", but gourmands have been enjoying them with eggs, onions and meat since Aztec times.

Matsutake

Tricholoma matsutake

*Japanese people place great emphasis on the seasons,
whether viewing cherry blossoms in spring or enjoying
snow on maple trees in winter.*

The matsutake harvest is a classic seasonal marker; the fungi are exchanged as gifts in September and October. A meaty, fragrant forest fungus smelling of spices, cinnamon and autumn, it was considered a noble food, eaten only by members of the Imperial Court in Kyoto. Women, it is said, were only allowed to refer to it by adding the honorific prefix "O" as a mark of respect: *O-matsu*.

Much as European nobility enjoyed "the chase", members of the Japanese court would go mushroom hunting to mark the seasons and, like many foragers today, were secretive about the best sites. Country folk, who already told rude jokes about the matsutake's phallic shape, started a tradition of stories about comic characters trying to hide their mushrooms (nudge, nudge), perhaps as consolation for being forbidden to eat matsutake themselves.

Thanks to their mycorrhizal relationship with mainly pine trees, matsutake are hard to cultivate commercially and must be found in the wild. They have been in decline since 1905, when Japanese forests became blighted with the pine weevil nematode, which kills the matsutake's host. However, the mushroom grows elsewhere, including the US and Canada As the market expands, foresters are increasingly seeing mushroom gathering as a way to bring alternative value and employment to former logging forests.

Matsutake are sold in seven grades, from grade one – large and firm, cap still closed, sealing in flavour and aroma – through to seven: a bit maggoty, but still worth having. Japanese law dictates the mushrooms are washed before sale, but many turn a blind eye to the rules so as not to impair the flavour. The highest prices are charged for native matsutake from the Tamba forest, near Kyoto.

Many people outside Japan are happy to make do with the *Tricholoma magnivelare* (white matsutake), which still has a fine flavour. *Tricholoma* has been used in medicine, and is an important food for musk deer, whose oil is prized by the perfume industry. Traditionally, however, this fungus has been so highly prized as a food that there is still much to discover about its potential therapeutic uses.

Opposite European matsutake (*Tricholoma caligatum*) considered edible but inferior to the Japanese matsutake.

N.º 21

E. M. Wakefield

Porcini

Boletus edulis

There is a reason why Boletus edulis *is sometimes known as the king bolete – it is one of the most sought-after edible fungi in the world.*

*E*dulis means "edible"; *Boletus* may come from the Greek *bolos* ("lump of clay") or from *boletaria*, the Roman cooking vessels used to prepare mushrooms. Thanks to its swollen stipe, *B. edulis* is known in France as *cèpe*, the Gascon word for "trunk". To British foragers, they are "penny buns", because one could easily be taken for a little bread roll, even occasionally sporting a "bloom" like a light dusting of flour. In Scandinavia, boletes are *karljohan svamp*, named for the monarch of a combined nineteenth-century Sweden and Norway, Karl XIV Johan. The fungus is prized wherever it is found. Fresh, the fruiting body makes for a hearty meal – so much so that they were a popular meat-free Lenten meal in Russia. Dried, it adds a distinctive, intense flavour year-round.

The ancient Romans called the mushroom *suillus* or "hog fungi", which is seen in the modern Italian porcini or "little pig", for its supposed resemblance to a chubby little brown piglet. Roman poet Juvenal believed they should be reserved for the gentry; everyone else should make do with less noble options. Pliny wrote of the bolete's potential medicinal uses; he noted they were good for complaints including bowel flux, haemorrhoids, freckles and dog bites. They have been used in folk medicine down the centuries, from Bohemian lumberjacks who ate them as a protection against cancer to foot-sore Latvians treating chilblains, while also helping with stomach ache and some heart conditions. In traditional Chinese medicine, boletes form part of traditional tendon-easing treatments.

Growing mainly in deciduous and coniferous forests, *Boletus edulis* enjoys the light, dappled shade of marginal woodland and clearings. Most boletes are ectomycorrhizal, meaning they share a relationship with the root systems of trees. Thanks to the fungi's mycelium, the tree roots absorb nutrients from the soil more easily; thanks to the tree, the fungi can absorb nutrients too.

Italian weather lore says porcini sprout with the new moon; other traditions claim bolete hunting is most successful under a full moon. Either way, the intricacies of timing are essential – harvest too soon, and the mushrooms will be tiny and less tasty; pick them too late, and they'll be full of maggots. Fruiting is triggered by rain after warm weather; a particularly wet autumn is known as a "bolete year".

Opposite Porcini or cep (*Boletus edulis*) by Kew mycologist Elsie M. Wakefield, Kew Collection, c.1915–45.

Bread and beer

To what shall I compare the kingdom of God? It is like leaven that a woman took and hid in three measures of flour, until it was all leavened.

Luke 13:20–21

The somewhat opaque "Parable of the Leaven" is generally thought to refer to how something very small (yeast) may be added to something large (flour) and cause it to grow in mysterious ways. In other words, it is the kingdom of God rising from humble beginnings.

Simple, single-celled and hundreds of millions of years old, there are over 1,500 species of yeast. Together, they form the oldest food "additive" we know.

In 2018, archaeologists found traces of the beer-making process in Raqefet Cave, Israel. Dating back 13,000 years, it toppled the previous oldest beer, a mere 5,000-year-old from Northern China. Bread may be 14,400 years old, though leavening – raising the dough using yeast – is much younger. While some scholars suggest the ancient Mesopotamians baked leavened bread, so far the earliest evidence is from Egypt, circa 1000 BCE.

Yeasts reproduce asexually via mitosis, where the cell splits into two identical copies of the original. As *Saccharomyces cerevisiae* (baker's or brewer's yeast) splits, it converts carbohydrates into carbon dioxide and, eventually, alcohol, through fermentation. This happens naturally – for example, via the sugars in ripe fruit, but when it is introduced to grains, the carbon dioxide acts to raise any bread or foam any ale made from them.

In beer, the yeasts work on the sugars in malt, or germinated barley grain. The first versions would have been left to acquire their own flowers from natural yeasts slowly multiplying in the atmosphere. Today the tradition continues in brews such as lambic beer from the Pajottenland region of Belgium, whose vats are left open to exploit the local wild yeasts. The result varies in each batch depending on the yeasts present; lambic can contain as many as 80 different species. The sheer number of variables makes it unsuitable for mass production.

Much like beer, the earliest bread would have been made from cultures allowed to develop via whichever wild yeasts were in the air. In 2020, the concept was reborn in the popular lockdown bread-making craze, using sourdough starters when baker's yeast was suddenly in short supply.

In Bulgaria, yeast bread held curative and protective properties because it would lose strength, but could be "renewed". The ritual of "storing the yeast" started on St Ignatius Day (20 December) and continued to the Feast of St Basil on 1 January.

Opposite Drawing of microscopic cells of baker's yeast (*Saccharomyces cerevisiae*).

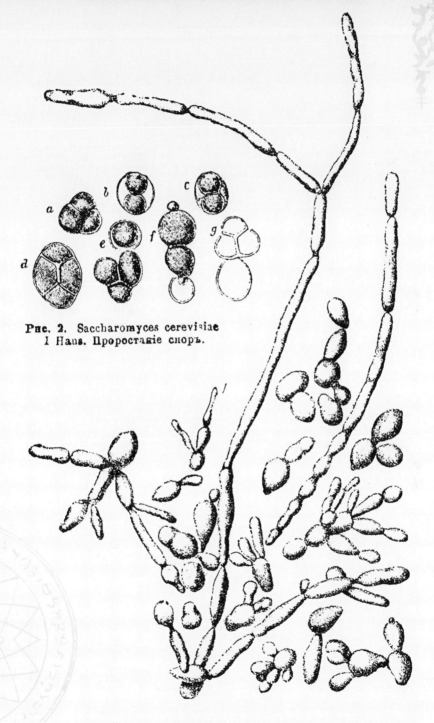

Рис. 2. Saccharomyces cerevisiae
I Hans. Проростаніе споръ.

Рис. 1. Saccharomyces cerevisiae I Hans. Клѣтки и мицеліевидныя
образованія изъ пленки на старой культурѣ.

Across Europe, *gebildbrot* (breads and pastries in various representative shapes) warded off evil on special occasions, including German pretzels, Belgian *speculaas* (spiced biscuits) and Swedish *Lussekatter* (saffron buns).

Bread in some form has been a staple of most cultures, and there are all manner of superstitions and customs surrounding the leavened loaf. For the Bedouin, a well-risen loaf implies a large family and wealth. In some Arabic communities, brides placed a piece of *hamíreh* (leavening) above the lintel of her groom's door to bring good fortune. In Turkey, yeast should never be given to a neighbour after *ezan* (call to prayer).

Above Print of a monk drinking with the devil, from Edith W Robinson *The Lay of Saint Jucundus: A Legend of York*, illustrated by George Hodgson. Nineteenth century.

Perhaps because bread was so important to life, many superstitions around it referred to death. If the dough cracked as it was being shaped, a funeral would soon be held. A large air bubble in a loaf also meant somebody would die soon, the hole representing their tomb (though other traditions thought it a sign that the woman who baked it was pregnant). In Yorkshire, if the loaf does not rise, there is a dead body nearby.

There are over 100 gods, goddesses, spirits, demons and heroes associated with beer and brewing, including Acan, the Mayan god of alcohol, Gabjauja, Lithuanian goddess of grain, and Bacchus/Dionysus, the Roman/Greek god of intoxication. Aegir, Norse deity of the sea, brews beer for the gods in a gigantic vat given to him by Thor, his best customer. At the other end of the supernatural hierarchy, *Biersal* are German house-spirit workers of the brewery. Every 28 February in Finland marks a celebration of the national epic poem *Kalevala*, which follows an ancient tradition of oral folklore. The poem includes an entire chapter on the brewing of beer, which takes twice as long to tell as the origins of mankind. Osmotar the brewer prepares and boils the mixture in the cauldron, then she allows it to "steep and seethe and bubble". The description reads as though it was a magical potion, and in some ways it was. No one knew why the barley and hops began to foam and, eventually, turn to alcohol.

Perhaps because people of ancient times didn't understand the link between yeast and beer, there are few deities specifically connected with fermentation. There is one candidate for "God of Yeast", however, the Lithuanian god Raugupatis, who breathed life into grain, causing it to rise, enabling bread and ale.

Morel family

Morchellaceae

When discussing the morel family, Pliny noted the resemblance of one honeycombed, hollow species to a sponge, and many do resemble a sponge expertly carved out of wood.

Many members of the Morchellaceae family are collected by humans, and are prized so highly that dedicated mycophiles, armed with special foraging sticks, closely guard locations of their quarry over a few precious weeks each spring. The family is associated with a rich, woodland flavour and smoky aroma, though Robert Rogers warns they must be cooked thoroughly, due to gastric irritants that are destroyed with heat, and in a well-ventilated space, allowing the toxic hydrazine in the fumes to escape.

Morchellaceae can look significantly different from one another, ranging in colour from pale cream to almost black, via honey yellow and dull brown. They can be tall and conical or short and rounded, but all are covered in deep pits and ridges, which range from deep ruffles to regimented rows of miniature bunkers. Pale cream or yellowish brown, *Morchella esculenta*, the common morel, enjoys chalky soil and light woodland. The black morel (*Morchella elata*) has a dark, conical cap on a white stipe. North American First Nations communities do not always agree on what they look like – the Mohawk call them "land-fish", while the Onondaga think they are more like penises, and the Cayuga think they resemble ears. Polish people have an earthier origin story: the devil was in a bad mood when he met an old woman in the woods, so he chopped her up and scattered the pieces across the forest floor, spawning the morel. In Russia, they are the *smorchki*, "princes among mushrooms". Russian peasants used to make bonfires in the woods to encourage the harvest, though the practice was banned as far back as 1753 thanks to the number of forest fires. One unexpected result of bombed-out houses in First World War France was an exceptional morel harvest for the neighbours. Rogers notes a bumper harvest the season after the famous 1980 eruption of Mount St Helens, in Washington state in the US. Some people are more fortunate, finding morels springing up in the woodchip mulch they bought for the garden.

A word of warning about the false morel, *Gyromitra esculenta*, which, despite its species name meaning "edible" in Latin, is not. A member of the entirely different Discinaceae family, it is still eaten in some parts of Europe, but can be extremely dangerous, causing damage to the kidneys, liver and central nervous system. False morel look more like wrinkled, dark brown brains than the pitted and ridged cones of the true morel. If in the slightest doubt, steer clear.

Opposite Morel (*Morchella esculenta*) from Anna Maria Hussey *Illustrations of British Mycology*, 1847–55.

Morchella esculenta, Dillon.

1796 Published by J. Sowerby London

Agaricus cantharellus

Chanterelle

Cantharellus cibarius

Russians know the chanterelle as lischki, *"little foxes". In Lithuania, they are "vixens", symbols of maidenly virtue, but this is unusual. Most traditions firmly associate chanterelles with the masculine virility of the cock.*

For many communities, the wavy edges of *Cantharellus cibarius* represent the showy comb of a cockerel. Certain Dutch mycophiles call it exactly that – the *hanekarn* or "cock's comb", while Italians know it as *gallinaccio* ("turkey cock"). Wasson points out the similarity between the French chanterelle with Chanticleer, the crowing cockerel of medieval folk tales. The name isn't universal, however. For First Nations folk in Colombia, the fungus is "little fish gills", to be eaten with salmon. In Germany, the name *pfifferlinge* refers to its slight peppery taste. Chanterelles are found in light forest floor and are mycorrhizal, growing in association with tree roots such as oak, hazel, sweet chestnut and birch. They can also spring up through the dried needle-beds of a pine forest. An eighteenth century legend claims that placing a chanterelle in the mouth of a dead man would return him to life. Other members of the Cantharellaceae family include the Pacific golden chanterelle (*Cantharellus formosus*), the state mushroom of Oregon. The red chanterelle (*C. cinnabarinus*) carries canthaxanthin, a carotenoid colouring found in the marine crustaceans that may turn their predators – flamingos – pink. It is sometimes used in suntanning agents.

Chanterelles are considered some of the finest edible mushrooms. According to Wasson, Norway and Sweden think of the chanterelle as the only worthwhile forest mushroom, though this attitude is rapidly changing. Indeed, foraging is becoming fashionable across many previously mushroom-hesitant nations. While this is generally a good thing, the sheer number of people disappearing into the woods with baskets (and, less sustainably, trucks) is beginning to threaten supply.

In order to preserve stocks, some countries or regions limit the amount of chanterelles – and other edible mushrooms – that people may gather. For example Haute-Savoie, in the French Alps, restricts individuals to half a kilo, or a kilo per car.

Even where foraging is allowed, learn from an expert before picking. The similar-looking woolly chanterelle (*Turbinellus floccosus*) can cause sickness in some. False chanterelles (*Hygrophoropsis aurantiaca*) and the darker orange jack-o'-lanterns (*Omphalotus olearius*) are both actively poisonous.

Opposite Chanterelle (*Cantharellus cibarius*) from James Sowerby *Coloured Figures of English Fungi or Mushrooms*, 1795–1815.

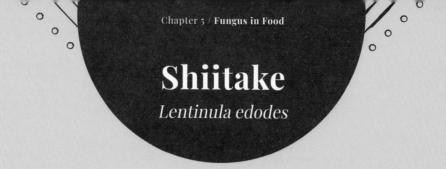

Shiitake

Lentinula edodes

*Also known as the "sawtooth", "black forest" and "golden oak" mushroom, shiitake is one of the most popular culinary mushrooms in the world, second in volume to the humble button (*Agaricus bisporus*), but infinitely superior in flavour.*

The name shiitake comes from its most regular native host, the shii tree – *Castanopsis*, a member of the beech family – and *take*, "mushroom". The Chinese name *xiang gu* means "fragrant mushroom". However it is known, it has formed part of Traditional Chinese and Japanese Kampo medicine for millennia, treating a number of conditions from cancers and melanomas to thread veins. Some cosmetics include shiitake extract for its cell-regeneration and skin-firming properties.

It has been known in Japan since at least 200 CE, when indigenous people are said to have presented shiitake to the emperor as an offering. Sometime around 1100, the legendary "Father of Shiitake", Wu San Kwung, a humble woodcutter, is said to have noticed the mushroom growing on logs he had recently felled, kick-starting an industry worth millions today.

Shiitake has a rich, smoky flavour that works with a range of dishes. It is best eaten cooked; happily, the process does not deplete the nutrients. Shiitake is a staple of Japanese cuisine, which divides it into two types. *Donko* is fat, rounded, and with a slightly open cap. *Koshin* is thinner, with a fully open cap.

Shiitake is not native to Europe, but it is simple to grow on freshly cut, medium-sized logs. Oak is considered best but most hardwoods will do. Spawn is sold in plug form, and is easily available online, along with any tools that make the drilling and sealing easier. To inoculate the logs, drill holes 7 to 10cm apart, fill them with shiitake spawn and reseal. The fresher the logs, the less chance there is of any contaminant fungi fighting the shiitake for supremacy. The logs should be stacked and covered to retain moisture and darkness. Once the mycelial growth has established, which can take up to three years, the logs can be forced into producing a crop by submerging them in cold water for 24 hours. Three to five days later, small mushrooms called "pins" will appear, which develop into mature mushrooms over a few days. After harvesting, the logs will need to rest for a couple of months before cropping again. Well-maintained log piles can produce four crops a year, for up to seven years.

Opposite Shiitake mushrooms (*Lentinula edodes*) illustrated by Malcolm English, 2022.

Fistulina hepatica. *With.*

Beefsteak fungus
Fistulina hepatica

Hollow trees make evocative settings for a folk tale. In the British Isles especially, lore often surrounds real trees, centuries old and garlanded with tales of Good King Henry or Good Queen Bess dancing around them with youthful exuberance.

Now hollow shells, these trees have seen it all. They have been used as boundaries, hiding places, prisons, pubs and even as the local hanging tree. Hollow trees were places of healing, where the sick might crawl, transferring their diseases to the tree. Yet without a certain fungus, these venerable giants would still have solid hearts of wood.

Bracket fungi hang off trees like miniature shelves or "conks", gaining sustenance from the woody centres of the trees that the fruit bodies grow from. *Fistulina hepatica*, or beefsteak fungus, is a typical example, consuming non-functional wood, most commonly from oak trees, and releasing locked up nutrition stored in the "dead" wood. The host tree also benefits by growing roots to tap into this compost-like resource that it could not create on its own.

The unlocked nutrients become home to a huge diversity of invertebrates and other organisms. The tree becomes hollow, but develops buttress roots, forming a lighter, stronger cylindrical structure, less likely to snap with age. Such trees are probably the most biodiverse microhabitats on land in Europe.

The beefsteak fungus is well-named – visually, at least. Looking remarkably like a lump of liver from the outside, the conk is cut open to reveal red "meat", marbled like fillet steak. It even "bleeds" – a red guttation drips from the fungus when young. While eaten by some, most agree it is tough and leathery. Its bitter taste grows stronger with age due to the high levels of ascorbic acid (vitamin C), which at least gives it nutritional value. Richard Mabey, King of Foragers, suggests dicing it and cooking it with strongly flavoured ingredients. Even then, he warns, it doesn't lose its sourness.

For something so redolent of fresh meat, there is little folklore surrounding the fungus itself, but any number of tales about its work: the ghostly white lady who haunts a hollow tree in the forest of Soeren in the Netherlands; the Scandinavian *hulder*, supernatural seducers found in the dark places of trees; the Skeleton Oak of Nant Gwrtheyrn, Wales, where a bride was lost on her wedding eve; and the Strangling Oak of Nannau Woods, where the Welsh revolutionary Owain Glyndŵr entombed his slain cousin Hywel Sele. Oh yes, without the beefsteak and its heart-rotting friends, folklore would be much the poorer.

Opposite Beefsteak fungus (*Fistulina hepatica*) from Anna Maria Hussey *Illustrations of British Mycology*, 1847–55.

Truffles

Tuber melanosporum,
Tuber magnatum

One of the most sought-after foods in the world, the "black diamond" has been prized by gourmands since antiquity, yet precisely why it holds such a fascination for humanity is still a matter of debate in the scientific community.

Tuber melanosporum, the rare black truffle, mainly found in France and Germany, is a common sight in comparison with its cousin *T. magnatum*, the white truffle, scarce even in its native Italy. Both fungi change hands for vast sums of money.

The *Tuber* genus includes mycorrhizal fungi that reward their hosts with an improved capacity to absorb nutrients. Famously growing under oak and hazelnut, the underground truffles are close to the surface, and carry a strong perfume to attract the woodland animals that will distribute its spores as they eat. Humans have long exploited this relationship, training dogs and pigs to hunt out the fungus, reining them back just in time to harvest the bounty. It is said young sows are the best truffle hunters of all.

The ancient Romans loved tuber almost as much as modern Italians; Juvenal was probably only half joking when he said it would be better for the grain harvest to fail than the tuber. The name gradually corrupted to the Italian *tartufo*, Spanish *trufa*, French *truffe* and English truffle. The word "trifle" purportedly comes from truffle too, corrupted 180 degrees to mean a frivolous item of little worth.

Truffle was a popular – if lavish – delicacy in Renaissance Europe, one that a 24-year-old John Evelyn experienced on 30 September 1644, in Vienne, France. He described it as "a certain earth-nut, found out by a hog trained to it ... an incomparable meat".

One of the most enduring folk beliefs about truffles is their reputation as an aphrodisiac. Recent research has shown that a compound in *Tuber* – alpha-androstenol – is also found in the saliva of rutting boars, and it has been suggested that this is why female pigs are so good at rootling out the fruiting bodies. Alpha-androstenol is also found in men's sweat and women's urine, though there has been no direct connection made with human sexuality yet.

Like so many fungi, truffles have fallen victim to deforestation, air pollution and overharvesting. While some truffles must be gathered, others can be farmed by planting orchards of oak and hazel trees previously inoculated with spores. Individuals may have a go at home by buying similarly prepared seedlings. Alas, success is not guaranteed.

Opposite Black truffle (*Tuber melanosporum*) from Anna Maria Hussey *Illustrations of British Mycology*, 1847–55.

Tuber cibarium. Sibth.

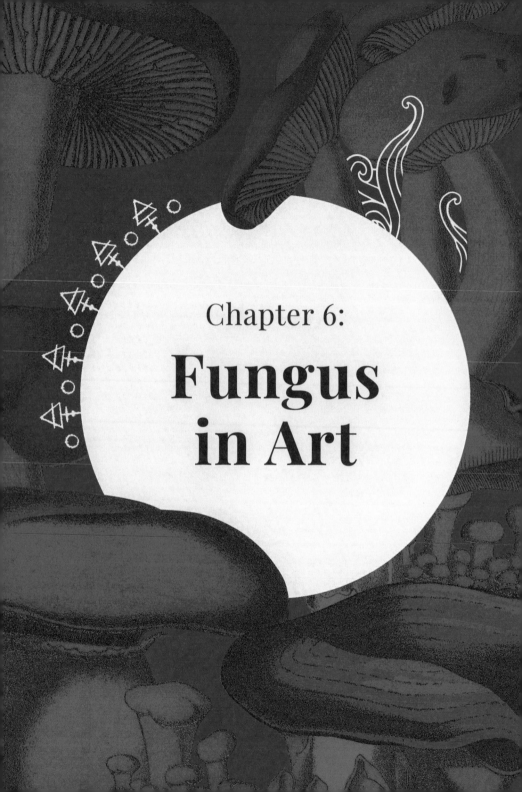

Chapter 6:

Fungus
in Art

There is something visually satisfying about a mushroom. Even countries considered traditionally mycophobic take endless pleasure depicting them, though their attitudes are rarely hard to read. Ancient Greeks, captured in the relief of a marble frieze, reverently offer one another a single mushroom. Roman mosaic floors and murals are filled with fungi – in dishes, bowls and baskets, spilling onto the cook's table, delivered to a gourmand's door. Early Christian art, however, shows mushrooms as evil, as the Tree of Knowledge – the downfall of Adam and Eve.

Art and Illustration

Valentina and Gordon Wasson consider the first depiction of a mushroom in modern art to be in *The Haywain*, a triptych by Hieronymus Bosch. In the central panel, next to a cart full of golden hay, surrounded by the usual Bosch mayhem of curious people and disturbing creatures, sits a huge *Boletus* mushroom. It sits proud on a little hillock, like a statue. The Wassons speculate that this is not the highly prized *Boletus edulis* (porcini), but *Rubroboletus satanas* or "devil's bolete". Their reasoning for this is not clear, as the Bosch mushroom does not have the characteristic red stipe of *R. satanas*, but it does fit in with early Christian attitudes to fungi.

The ambivalence continues with the centuries. Luscious seventeenth-century still lifes with mushrooms, for example, appear by Italian artists such as Simone del Tintore and an anonymous Spanish artist, both from mycophilic countries. But at the same time, Otto Marseus van Schrieck depicts fungus as an agent of decay. *Still Life with Insects and Amphibians* (1662) shows mushrooms alongside snakes, moths, toads and, perhaps most telling of all for a Dutch artist, a dead tulip. Thomas Gainsborough's 1785 *Mushroom Girl* holds a basket of mushrooms as a prop, rather than as something she may have just gathered with any relish. Gordon Wasson calls it "an excuse for a title", though he grudgingly admits the existence of another Gainsborough that depicts a child reaching eagerly for a mushroom.

From the eighteenth century, as mushrooms began to be associated with the fairy folk, they began to get cuter. *The Puck,* as painted by Joshua Reynolds in 1789, shows a cherubic Puck sitting on a toadstool clutching a posy, though a certain glint in his eye tells us this little *putto* ("naked child") may not be as angelic as most. Equally ambiguous is the work of Richard Dadd, whose 1841 depiction of the same character is similar in pose to Reynolds's, but with a bunch of cavorting fairies under the mushroom, lending a more sinister feel. Dadd's 1842 depiction of Ariel's fairy ring in *The Tempest, Come Unto These Yellow Sands*, is positively apocalyptic. Red-tinged, crazed dancers cavort around the arch of an ancient rock, against a blackened sky and foaming sea. On the eerily lit sand lies an abandoned mortar, its accompanying pestle looking distinctly like a flaccid *Phallus impudicus*, the stinkhorn mushroom.

Above Stocking webcap (*Cortinarius torvus*) by
Beatrix Potter.

Opposite Illustration by Arthur Rackham of a fairy
ring for *A Midsummer Night's Dream*, 1908.

Literature and Film

Upon a mushroom there is spread
A cover fine of spiders web.

The Pastime of the Queen of the Fairies, Margaret Cavendish (1623–1673)

As the Victorian age progressed, the tone changed. Fairies shrank from sinister, supernatural beings into twee "little folk" that lived in mushrooms. John Anster Fitzgerald's *The Intruder* (c.1860) is typical of the genre. A party of beautiful, wispy fairies (and a couple of impotently angry goblins) confront a warty but ultimately unthreatening toad who has wandered under their *Amanita muscaria* (fly agaric).

The Victorian craze for nature now saw botanical illustration as a serious art form. Artists, often women, including Beatrix Potter, painted fungi in meticulous detail. This freed up artistic representation to be used in a more fantastical sense. Another nineteenth-century fascination, the cult of childhood (in the middle-class sense, of course; working-class children were still sent up chimneys) saw a proliferation of fairy stories and books for small people, infantilizing mushrooms into nursery bogeys. Richard Doyle's *In Fairyland: A Series of Pictures from the Elf-World* (1870) is typical of art incorporating mushrooms into whimsical images of sprites and elves playing with nature. Even when mushrooms did not take an active role in fairy tales, they became a constant part of the illustrations. The endlessly pictorial fly agaric acted

as seats for delicate-winged creatures, homes for gnomes, umbrellas, wheels and street-market wares, filling gaps in backgrounds and decorated borders. Artists who included them, such as Arthur Rackham, Charles Heath Robinson, Cicely Mary Barker, Hilda Boswell and Molly Brett, are still loved by millions.

The twentieth century saw another dive back into ambiguity as art explored mushrooms in a wider sense. A gigantic mosaic of a stinkhorn erupted from what looks like Hansel and Gretel's gingerbread house in Antoni Gaudí's "mushroom house" at the entrance to Barcelona's Parc Güell, built between 1900 and 1914. The pop artists of the 1960s would have been only too aware of the psychedelic effects of *Psilocybin* mushrooms, producing all manner of curious interpretations. Darkest of all must be Roy Lichtenstein's 1965 Atom Burst, a mushroom in cloud form, anything but cuteness on its mind.

In her book *The Pastime of the Queen of the Fairies*, written sometime in the mid-seventeenth century and showing Queen Mab using a toadstool as a table, Margaret Cavendish was continuing a literary tradition. Equating fungi with the fairy folk was made popular by the likes of William Shakespeare, whose Prospero notes that elves "make midnight mushrooms" in *The Tempest*.

When it comes to fungi in literature, however, there is only one "big daddy": Lewis Carroll. In

Opposite "Advice From A Caterpillar", illustration by Sir John Tenniel from the first edition of Lewis Carroll's *Alice's Adventures in Wonderland*, 1865.

one of the most famous scenes in literature, Alice encounters a caterpillar smoking on a giant toadstool. He invites her to try a little of the mushroom; one side will make her grow taller, the other will make her shrink. Taking a piece from each side, Alice accidentally shoots to enormous heights, descends into microscopic size, then learns to regulate the amount she eats for each situation as she negotiates her way through Wonderland. The idea has been the blueprint for any number of similar themes, across genres; even Japanese games designer Shigeru Miyamoto has acknowledged the direct influence of Lewis Carroll in his creation, Nintendo's *Super Mario*. In the game, Mario or Luigi can eat mushrooms to either grow or shrink in size.

We will never know whether Charles Dodgson (Carroll's real name) was curious about the hallucinogenic side of fungi when writing *Alice in Wonderland*. The Victorians were certainly into various mind-altering drugs, opium addiction being one of the main social problems of the day. Passionate arguments rage on both sides, though it is not likely that the non-smoking Reverend Dodgson, a moderate drinker who was unimpressed with narcotics such as opium, would have sampled the dubious delights of *Amanita muscaria* for himself. However, he may well have read about it, as various scholars have pointed out. Historian Michael Carmichael has pointed out that a few days before Dodgson began writing *Alice in Wonderland*, he made his one and only visit to Oxford's Bodleian Library, where Mordecai Cubitt Cooke's 1860 drug survey *The Seven Sisters of Sleep* had just been deposited. Carmichael reminds us that Dodgson had seven sisters and was a lifelong insomniac. He also reveals that most of the pages of this particular copy remained uncut when he examined it, with one exception: the chapter about fly agaric.

The Wassons note that Cooke brought out another book, *A Plain and Easy Account of British Fungi*, in summer 1862, where he discusses the Koryak people of Kamchatka, Russia, and their use of fly agaric as a hallucinogen. Dodgson sat down to write his novel on 13 November of that year. Then again, perhaps the publication dates are just coincidence. Perhaps he went to the library to read the newspaper and his caterpillar and magic mushroom are mere figments of his imagination. The sketch he drew to accompany the scene does not depict *Amanita muscaria*, but a simple, plain, flat-topped fungus. He certainly couldn't have got the idea from Grimm. The line in Jacob Grimm's *Teutonic Mythology*, "whoever carries a toadstool about him grows small and light as an elf", didn't appear until 1888. In a world full of trippy mushrooms and pop art, perhaps we are reading too much into a simple children's story.

The Wassons are surer of Mordecai Cubitt Cooke's influence on Charles Kingsley in his 1866 novel *Hereward the Wake*, in which a nurse from Lapland adds scarlet toadstool juice to men's ale so she can discover their secrets. Valentina Wasson finds more tender associations with mushrooms in Tolstoy's *Anna Karenina*, which associates mushroom gathering with love, family and togetherness.

Wasson points out that only in a mycophile country like Russia would such scenes be found, and in many ways, she is right. Most Western depictions of mushrooms take a much darker turn. English writer William Hope Hodgson must have had a bad experience with mushrooms. His short story "The Voice in the Night", published in November 1907 in *Blue Book* magazine, tells the story of a shipwrecked man and his fiancée, lost in a mysterious lagoon. The island they believed would save them is inhabited by

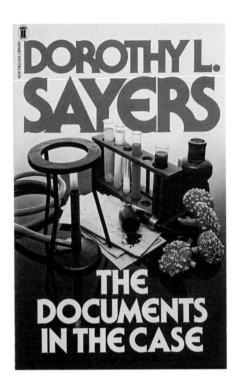

DOROTHY L.
SAYERS

THE
DOCUMENTS
IN THE CASE

Opposite
*The Documents in
the Case* by Dorothy
L. Sayers, originally
published in 1930.

a strange fungus that gradually spreads and absorbs them. The story was most famously adapted in 1963 as *Matango*, a Japanese horror film, released in English as *Attack of the Mushroom People*.

Somehow, fungus invasion seems to suit the short-fiction genre. John Wyndham's *The Puff-Ball Menace* was originally published in 1933, in which the dastardly warmongers of "Gangistan", knowing they cannot beat the might of Britain, invent a biological weapon in the form of puffball mushrooms. Wyndham was writing for *Wonder Stories* magazine as John B. Harris and, under the better-titled *Spheres of Hell*, his tale tapped into the "yellow peril" racial stereotypes of the day. Ray Bradbury's 1966 collection *S is for Space* features the story "Come into My Cellar", where a boy growing mushrooms in his cellar may be

inadvertently farming aliens. Bradbury originally wrote the story in 1959 as a genuinely creepy TV script "Special Delivery", for *Alfred Hitchcock Presents*. It was remade in 1989 as a comedy horror, *Boys! Raise Giant Mushrooms in Your Cellar!* and has also featured in a 1992 Dave Gibbons comic strip. Until recently, however, fungus has not figured heavily in movies, perhaps due to the West's traditional reticence to understand how mycology works. This is slowly changing, though perhaps not the way mycologists might hope. *Shrooms* is a 2007 slasher movie about teenagers stalked while foraging magic mushrooms, while *In the Earth* is a 2021 horror film exploring the ways mycorrhizae can go bad. Oh, for the innocent days of happy, blobby mushrooms dancing to the *Nutcracker Suite* in Walt Disney's 1940 film *Fantasia*.

Whodunnit? Fungus in crime fiction

It is a truth universally acknowledged that a crime writer in possession of a dodgy mushroom must be in want of a victim. Yet fungi are more rarely used in whodunnits than might be supposed. Warning: spoilers ahead.

In 1914, English author Ernest Bramah (1868–1942) created Max Carrados, a blind detective who, in his day, was ranked alongside Sherlock Holmes. In Bramah's *The Mystery of the Poisoned Dish of Mushrooms* (1923), Carrados investigates the ability of a fictional toadstool, *Amanita bhuroides*, to kill within half an hour of ingestion. In H.G. Wells's *The Purple Pileus*, his suicidal protagonist consumes a poisonous-looking toadstool, believing it to be deadly. Wells's mushroom needed properties that *Amanita muscaria* (fly agaric) doesn't have, so he, too, invented a new one.

Perhaps crime writers tend to avoid real mushroom poisoning because it is so easy to get the details wrong. Gordon Wasson was scathing about crime writers perverting his beloved fungi for their own evil ends. Writers might, he suggested, make their victim die too quickly, as with the 1950 *Murder with Mushrooms*, by Gordon Ashe (a pen name of John Creasey). Perhaps they'd use an unreliable species, such as August Derleth's 1948 short story, named merely "?", in which the murderer substitutes a false morel for the real thing. Wasson was unconvinced, considering *Gyromitra esculenta* not reliable enough to kill. Wasson did reserve praise

for Anne Parrish who, in 1925, did the legwork for *The Perennial Bachelor* and researched *Amanita phalloides* to his satisfaction.

Dorothy L. Sayers's *The Documents in the Case* revolves around a synthesized version of muscarine, a substance found in fly agaric. Muscarine is destroyed by cooking, Wasson points out, so could not have killed the victim.

The most famous crime writer of all, Agatha Christie, had a passion for unusual toxins, but rarely used fungi, although her 1928 short story The Thumb Mark of St Peter does see an innocent person accused of murder by mushroom. 1939's *Murder is Easy* did not have the murderer "adding to" wild mushrooms simmering in a pot; that particular twist was added for the 2009 TV series *Agatha Christie's Miss Marple*. A more recent TV detective does find use for fungal poisonings. At least three of Dr Gregory House's mystery ailments in the medical procedural *House* turn out to have fungal perpetrators. Perhaps the mushroom as murder weapon will see its day in court yet.

Opposite Growth stages of sporing body (mushroom) of death cap (*Amanita phalloides*) by Elsie M. Wakefield, Kew Collection, c.1915–45.

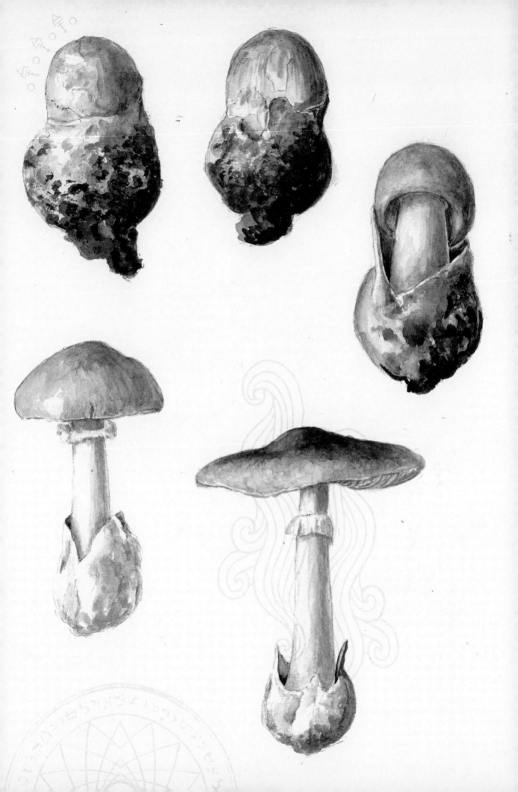

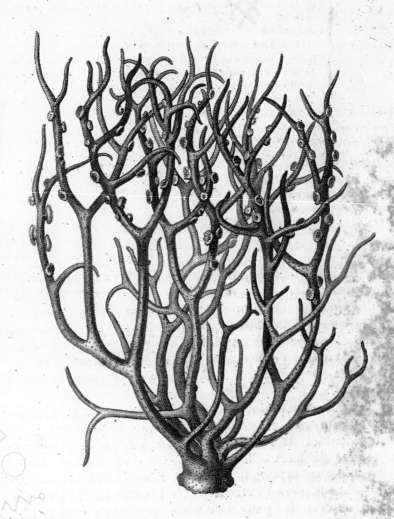

Fungi to dye for

Once the height of fashion, mushroom dyes fell out of style in the nineteenth century only to make a glamorous comeback a century later.

Lichens have an extraordinary ability to change according to the acidity in their surroundings. This was noticed by Spanish physician Arnaldus de Villa Nova sometime around 1300, when he discovered litmus (from the old Norse word *litmosi*, literally "dye-moss"), a way of testing the pH balance of materials. However, he wasn't the first person to put the idea to practical use.

Purple was the colour of power, of the emperors of Rome and beyond. Tyrian purple, made by the people of Tyre, was pounded from marine molluscs, but as those became overfished, people looked elsewhere. Ezekiel 27:7, in the Old Testament, describes "awnings of blue and purple" from the coast of Elishah (Asia Minor). These would have been dyed with *oricello*, made from the *Roccella tinctoria* lichen. The process involved an unappealing mixture of lichen, stale urine and slaked lime. Adjusting the species of lichen and the alkaline content created different shades. Lichen dyes do not require a mordant (fixing agent), but the extra colours that could be obtained from other fungi made experimentation with mordants such as alum, iron and copper sulphate worthwhile.

Dead man's foot (*Pisolithus arhizus*) redeems itself when called by another common name, dyemaker's puffball. It may look like an unpromising lump of dung, but it yields yellow, brown and even gold dyes. The fabulously frilled dyer's polypore (*Phaeolus schweinitzii*) is bright yellow, but can turn wool and silk a satisfying olive-brown. Crottle (*Parmelia saxatilis*) makes the deep rust-browns of Harris Tweed.

In the nineteenth century, the tradition of using natural resources to colour textiles waned as artificial dyes took over. Molluscs and endangered lichen everywhere could rest easy as synthetic mauveine took on the responsibility for purple. Only in the mid-twentieth century did certain craftspeople, led by American artist Miriam C. Rice, begin to re-learn ttaditional techniques. Rice experimented, mixing and matching with modern techniques and sharing her results with other like-minded individuals. Her book *Let's Try Mushrooms for Color* (1974) brought the concept to a new generation and, in 1985, the International Mushroom Dye Institute was born.

Rice's many books reveal various mushrooms and the mordants required for individual colours. Keen fungus-dyers can attend one of the famous International Fungi and Fibre Symposia, though as I write, the seventeenth such festival has been sold out for months. Fungal dyes are truly back in style.

Opposite Lichen (*Roccella tinctoria*) from F.P. Chaumeton *Flore Médicale Décrite*, 1815–20.

Foxfire

About 2,300 years ago, Aristotle noticed a "cold fire" of fungi illuminating a forest by night, and wondered how that could be. He never came up with a good answer; scientists have been working on the "foxfire" question ever since.

Many organisms emit luminescence, from glow-worms to marine algae, but among the least understood are the 71 species of fungus – out of around three million – that emit an eerie green light. Folklore was as puzzled as the scientists, but assumed supernatural and possibly evil reasons for these "fairy lanterns".

The phenomenon was sometimes known as will-o'-the-wisp, said to lure humans to their deaths, but the term is confusing because it is also applied to the light produced when marsh gas ignites. More specific to fungi is foxfire, perhaps named for the presence of foxes in the woods where it occurs, or a corruption of the French *faux* or Old English *fols* – "false". Californian miners saw rotten timber struts colonized by luminous fungi as "jack-o'-lanterns", covered in "timber frost" in places where their predecessors had died. In Utah, a glowing, ghostly "hair" (mycelium) marked where a miner had perished. Missouri's "spook light" legend may have been *Panellus stipticus* (bitter oyster), which grows around birch, oak and beech trees.

Some bioluminescent species are widely found; others are unique. *Mycena lux-coeli*, the heavenly light mushroom, is found only in forests and mountainous regions of Japan. Only the stipe of the eternal light mushroom (*Mycena luxaeterna*) glows by night, but the *flor de coco* ("coconut flower") *Neonothopanus gardneri* is said to be bright enough to be mistaken for fireflies.

Nineteenth-century French chemist Raphaël Dubois discovered that the glow is produced by oxidative enzymes. Fungi control the light emission using a circadian clock. In 2015, US and Brazilian scientists finally cracked Aristotle's foxfire conundrum: why did fungi glow? The team fashioned a set of fake, LED-lit mushrooms, then watched as they attracted all manner of insects. They now believe glow-by-night fungi literally light the way for the creatures who, in return for a meal, distribute spores for the next generation of ghostly mushrooms.

Don't be fooled. The incredible photographs on the internet can make bioluminescent fungi seem much brighter than they are. For our ancestors, unused to electric light, they would have been easier to spot, and there are accounts in various cultures of their being used as improvised torches. The best way to view bioluminescent fungi with our light-polluted eyes is to identify the mushroom in daylight and return on a moonless night, slowly adjusting your vision.

Opposite Bitter oyster fungus (*Panellus stipticus*), known to be bioluminescent in North America. From Anna Maria Hussey *Illustrations of British Mycology*, 1847–55.

Chapter 7:

Flying High

The association between sorcery, "flying" and fungi is well-documented. In Austria, fly agaric (*Amanita muscaria*) is even known as the witch's mushroom. Yet the psychological effects of fungi have been known and exploited by humans for as long as we have been around and also, perhaps, by animals for even longer than that.

The *Rig Veda* is one of the oldest devotional texts known to humanity. Written in Sanskrit, it consists of a series of Vedic hymns dating to sometime in the second millennium BCE.

It contains several references to a mysterious ritual drink called *soma*, made by extracting and fermenting the juice of a certain plant. Scholars have pondered the identity of this "god plant" for centuries, suggesting everything from somalata (*Cynanchum acidum*) to electrum, a naturally occurring alloy of gold and silver. The Wassons suggest *Amanita muscaria* as the most likely option, though others have proposed a number of different mushrooms containing psilocybin. One thing is for sure: mushrooms can do strange things to the fauna of this world.

Ethnomycologists have made extensive study into the use of psychoactive fungi in Mesoamerica, including the mysterious mushroom cult that carved strange stones depicting fungi growing out of human backs. Only about two hundred of these stones remain; most were smashed by horrified sixteenth- and seventeenth-century missionaries. But even as they destroyed the evidence, the missionaries created more. Their accounts of rituals by the "new" kids on the block, the Aztecs, form the bulk of what is known about historical Mexican mushroom rituals today. Much of the second volume of the Wassons' *Mushrooms, Russia and History* analyzes early sources for psychoactive mushrooms, including observations by Spanish priests.

Some talk of the great prince Montezuma entertaining his enemies, the Tlascalan princes, with "inebriating mushrooms". Others describe ritual sacrifices, where both victims and onlookers ate raw mushrooms until they became more inebriated than if they had drunk wine. Dominican friar Diego Durán suggested users became so intoxicated that the devil spoke to them in visions; a few even took their own lives. Franciscan priest Toribio de Benavente Motolinía was shocked that some of these pagan rituals seemed to resemble a horrific parody of Holy Communion, with the mushrooms representing the Host (sacramental bread). The Aztecs, he said, even called the mushrooms "god's flesh". Consumed with honey, the fungi invoked "a thousand visions" and users felt as though worms were eating them alive. Another of the Wassons' "witnesses", Francisco Hernández, was a botanist who travelled to Mexico in 1570 to study New World flora. He spoke of three species of "sacred" mushroom. One conjured war and demons; a second, *teyhuinti*, incited uncontrollable laughter. An unnamed third was so rare that people sought it in all-night vigils. Alas, Hernández's original manuscript was lost in a fire, so it's impossible to know for sure what these fungi were.

The most famous of the missionaries, Franciscan Bernardino de Sahagún, studied the customs and culture of the Mexican people for 61 years, between 1529 and 1590. His *Historia General de las Cosas de Nueva Espana* is written in both Spanish and Nahuatl, often side by side. He describes *nandcatl* mushrooms as small and black, consumed at special gatherings

Opposite Various stinkhorn mushrooms (basimycetes) by Ernst Haeckel, 1904.

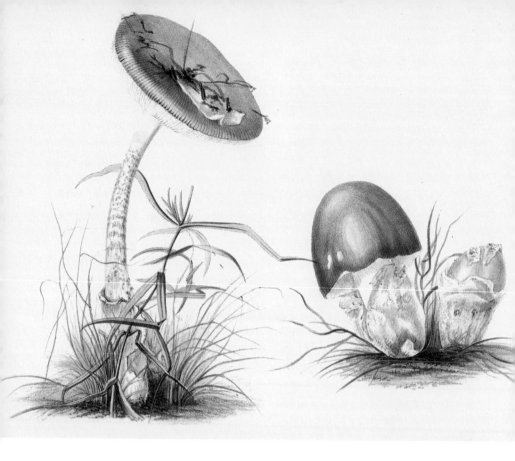

with varying psychological effects, though again it is uncertain what this mushroom may have been.

A great number of fungi have psychoactive effects, and they are by no means all identified. For example, the fool's mushroom is, today, associated with *Amanita virosa*, the destroying angel (see page 169), but the name is more ambiguous in previous centuries. The sixteenth-century botanist Charles de l'Ecluse (better known as Carolus Clusius) describes a mushroom in Germany as *narrenschwamm*, meaning "foolish fungus" or "fungus of fools", because "he who eats [it] becomes mentally upset". The Wassons tell us that if someone is acting weirdly in Hungary they are still asked if they've been eating *bolondgomba* (fool's mushroom). Viennese people accuse a neighbour acting strangely as having been on the "mad mushrooms". Similar sayings arise in Poland and Slovakia. Since destroying angel is so toxic that it will cut to the chase and just kill the consumer, this fool's fungus must refer to a different psychoactive mushroom. Clusius suggests *Amanita vaginata*. The Wassons suspect it was more likely a member of the *Panaeolus* genus.

In 1916, American surgeon Dr Bearman Douglass, his wife and maid ate some mushrooms on toast. An hour later, they experienced loss of balance, silly talk, uncontrolled laughter and a desire

Above: Grisette mushroom (*Amanita vaginata*) from Anna Maria Hussey *Illustrations of British Mycology*, 1847–55.

to make loud noises. Their minds swam, nearby objects appeared miles away and everything seemed diminished in both size and volume. Six hours later, everything was back to normal. The Wassons believed the symptoms to be the result of eating *Panaeolus papilionaceus* (petticoat mottlegill) even though it is known to induce nausea, something the Douglass family did not experience. Further research is required – only fools need apply.

During the Age of Enlightenment, scientists started to try to make sense of how humans had used "magic" mushrooms throughout history. In 1784, Samuel Ödmann, a Swedish priest and all-round naturalist, began thinking about Norse uses of hallucinogenic mushrooms in his famous work *An Attempt to Explain the Berserk-raging of Ancient Nordic Warriors Through Natural History*. Ödmann suggested that the legendary Viking crack squad the Berserkers (literally "bear shirts") consumed *Amanita muscaria* before going to war. Tanked up, hallucinating and mad as Valhalla, they raged across the battlefield cutting down their enemies like stalks of corn. Although the idea of consuming magical fungi fits the image of crazed, super-strength warriors, there is no actual evidence for the theory, archaeological or literary, and until there is, Ödmann's must remain a deliciously seductive theory.

Similarly, the legendary character of Bǫðvarr Bjarki, in the *Saga of Hrólfr Kraki*, has been linked with the hallucinogenic properties of fungi. The Norse shape-shifter always seemed to be asleep while his companions were fighting, though an other-worldly spirit in the form of a monstrous bear wreaked havoc against Bjarki's enemies on the battlefield. Bjarki was awakened and told to join the fray. The bear disappeared and the battle was lost.

The 1950s and 1960s saw an explosion in experimentation with the hallucinogenic properties of fungus, kick-started by the likes of Timothy Leary and his Harvard Psilocybin Project (see page 125), leading to the psychedelic hippy subculture of the 1960s and 1970s. Today, scientists continue to probe, but also make discoveries while looking for practical solutions to everyday problems. Researchers at Clarkson University in New York, studying ways to improve air quality in regular houses, have suggested possible connections between the inhalation of mould spores and sightings of "ghosts". One example may be *Stachybotrys chartarum*, or toxic black mould, found in "haunted houses" – and dirty bathrooms – across the world.

Above: USSR-era Russian postage stamp depicting fly agaric (*Amanita muscaria*), 1986.

Magic mushrooms

Psilocybe

When Samuel Taylor Coleridge noted the remarkable resemblance between the famous eighteenth-century symbol of freedom, the cap of liberty, and Psilocybe, it was a random association – he does not seem to have been aware of the mushroom's now famous properties of "magical liberation".

Coleridge noted that the fungus's conical cap and long, slender stem looked just like the Phrygian bonnets carried on poles as symbols by freedom movements as diverse as William of Orange's Glorious Revolution, the American War of Independence and the French Revolution of 1789.

"Magic" mushrooms – which can include most members of *Psilocybe* – appear in ancient Egyptian tomb paintings, Scandinavian myths, medieval illuminations and even a rediscovered first-century Mongolian textile. This extraordinary fragment, possibly depicting the brewing and consumption of the mythical ritual drink *soma*, was found at the burial site Noin-Ula in northern Mongolia. Alongside the warriors, the king and an altar, a priest holds the "divine mushroom" over a fire.

Psilocybe have varying amounts of the compounds psilocybin, psilocin and baeocystin. One or two do not contain any at all and have been moved to a new genus, *Deconica*. Similarly, of the approximately 180 varieties of fungus that contain psilocybin and psilocin, not all are in the *Psilocybe* genus.

The genus name comes from the Greek for "bare head". Many of the group derive both common and scientific names from their habit of using dung as a growing medium. *Deconica coprophila*, the round dung mushroom – previously thought to be in the genus *Psilocybe* but now *Deconica* – translates as "faeces lover". *Stropharia stercoraria* (dung roundhead) is named for Sterculius, the deity at the back of the queue when ancient Roman patronages were being given out, being god of manure, foul odours and the privy. Confusingly, this particular mushroom contains no psilocybin, so has now moved to the new *Deconia* genus.

Even among the mushrooms that contain psychoactive compounds, the results are not always what would-be drug dabblers might hope for. Alongside the dizziness, hallucinations, weird hilarity, space and time distortion and general "stoned" effects, users can experience headache, bone ache, acute depression, vomiting and ataxia (problems with coordination, balance and speed).

One of the earliest accounts of the use of psilocybins is in the Icelandic *Saga of Erik the Red*. Þorbjörg, one of the *lítilvölva* seer sisters, uses a "magical substance" to descend into a trance and

Opposite Psychedelic wavy cap mushroom (*Psilocybe cyanescens*) by Elsie M. Wakefield, Kew Collection, c.1915–45.

pains and numbness in their fingers and toes. J.S. staggered into the street to get help, but became dazed and confused. The physician Everard Brande was summoned. He was so curious at the sight of the family staggering around that he made precise notes of the symptoms, which he then published in *The Medical and Physical Journal*. Dilated pupils, racing pulses and breathing difficulties came in waves, along with a hysterical fear of dying that attacked all but the youngest family member, eight-year-old Edward S., who couldn't stop laughing. He "spoke nonsense", appearing not to know or care where he was or what he was being asked.

Dr Brande attributed the poison to fungus, but accused the well-known fly agaric (*Amanita muscaria*). Today, the symptoms suggest *Psilocybe semilanceata* (liberty cap), giving J.S.'s family the first-recorded magic mushroom trip. J.S. was unusual enough to tempt a visit from botanical illustrator James Sowerby, who was working on his magnum opus *Coloured Figures of English Fungi or Mushrooms* (1795–1815). Sowerby asked J.S. to describe the fungus; his illustrations leave us in no doubt that the breakfast consisted of liberty cap, along with another since identified as being from the related *Stropharia* genus.

Despite widespread knowledge of the incident and a flourishing underground Victorian drug culture, it would take until the 1950s to isolate the psychedelic compounds within the fungus. Albert Hoffman, who had earlier discovered LSD, began investigating hallucinogenic mushrooms in the 1950s, isolating psilocybin in 1953. He found it in the liberty cap in

predict the future. Yet somewhere along the way, the hallucinogenic properties of psilocybin seem to have become suppressed in the general European mind – or at least in most recorded accounts. Certainly by October 1799, when a chap known only as J.S. gathered mushrooms in Green Park, London, and put them in a stew, the results were surprising enough to log in an academic journal. Soon after, J.S. started seeing spots and colour flashes and found it hard to stand. His family had stomach

Above Victorian photograph of a woman posing beneath a giant model toadstool.

Opposite *Okame Laughing at the Shadow of a Mushroom*, wood-block print by Yoshitoshi, 1882.

1963. The samples were given to him by the Wassons, who identified several *Psilocybe* species, including *P. mexicana* (Mexican plum), *P. caerulescens* (landslide mushroom) and the classic magic mushroom, *P. cubensis*. The Harvard Psilocybin Project ran between 1960 and 1962. It was led by Dr Timothy Leary, a clinical psychologist whose experiments with magic mushrooms, and other psychedelic drugs like LSD, are legendary. Leary's work – and, indeed, the entire project – was highly controversial, both ethically (due to his choice of test subjects) and legally, but it would become the bedrock of the hippy counterculture. Psilocybin is currently being explored once again as a treatment for mental health conditions such as PTSD, depression and addiction. Just as bird watchers use the acronym LBJ (little brown job) for a number of small brown birds with indistinct markings, mycologists sometimes refer to LBMs: little brown mushrooms, for several unobtrusive common species, many of which are highly poisonous. Many *Psilocybe* species are LBMs, making identification challenging. Don't take a chance. Walk away from the little brown mushroom.

Under the UK Misuse of Drugs Act 1971, psilocin or a psilocin ester (any compound derived from it) is considered a Class A, controlled drug. The 2005 Amendment specifically includes fresh "magic mushrooms" in the Class A category.

Agaricus aureus, *Bull.*

Spectacular rustgill

Gymnopilus junonius

The Konjaku Monogatarishū (Anthology of Tales from the Past), *written in Japan during the Heian period (794–1185), consists of over 1,000 stories in 31 volumes, 28 of which survive.*

One legend tells how some woodcutters from Kyoto became lost in the forest. After wandering for some time, they came across four or five Buddhist nuns behaving strangely. Dancing, singing and laughing, they told the men how they, too, had lost their way – but had found some mushrooms to stave off their hunger. They offered to share the fungi with the woodcutters, and soon they were all in fits of hilarity. Ever since, the mushrooms have been known as *maitake* (dancing mushrooms) or *warataike* (laughing mushrooms). Laughing mushrooms turn up in other folk tales and in the names of several traditional Japanese dances, both comic and serious.

It's not entirely clear which mushrooms the jolly fairy-tale nuns actually found. There are around 30 species containing hallucinogens in Japan and several are known as laughing mushrooms, including *Gymnopilus junonius*, *Panaeolus papilionaceus* (petticoat mottlegill, said to be beloved of Portuguese witches) and *Psilocybe subcaerulipes*. Often known in the West as "laughing Jim" or "fiery agaric", *Gymnopilus junonius* grows on dead wood, clumping around the base of old trees, and can grow quite large – caps are usually between 7 and 20cm across, orange or red-brown in colour. It is sometimes mistaken for honey mushroom, though it changes colour to an unappetizing green upon being cooked, and tastes unpleasant. David Arora, quoted by Robert Rogers, says they are not hallucinogenic, though the equally bitter blue-green *G. aeruginosus* does have some such properties and is also found in Japan. *G. junonius*, aka the laughing gym, is sometimes equated with an ancient Greek fungus found on oak trees that was thought to facilitate clairvoyance. Robert Rogers says *G. junonius* is considered edible in parts of Europe but, even as a confirmed mycophile himself, confesses: "I wouldn't try it."

It is possible magic mushrooms of some description were used by the Jōmon people of prehistoric Japan, 10,500–300 BCE. A number of ceramic mushrooms have been found among many ritualistic materials excavated in the Akita prefecture.

Like many fungi, *Gymnopilus junonius* is being eagerly investigated by scientists for possible therapeutic effects, and is showing encouraging results against sarcoma and carcinoma. Its relation *G. liquiritiae* is revealing similarly exciting possibilities. The magical kingdom of mushrooms still has much to teach – and perhaps give to – the human race.

Opposite Rustgill, laughing gym (*Gymnopilus junonius*) from Anna Maria Hussey *Illustrations of British Mycology*, 1847–55.

Flying
ointment

Our image of a witch today is usually that of an ancient crone,
complete with pointy hat, warty nose and magic broomstick.

In the Middle Ages, however, a witch might be foul or fair, old or young, female or male.

The Catholic Church was suspicious of anything it didn't have direct control over. It suited the Church's propaganda machine to depict individuals who did not live a conventional existence in as depraved a manner as possible — and what could be more damning evidence of consorting with the devil than attempting to fly?

"Witches" were shown flying on broomsticks as early as 1451, when the woodcut *Hexenflug der Vaudoises* (Flight of the Witches) was taken more seriously than the satirical poem it illustrated, "Le Champion des Dames" (Ladies' Champion). It had serious consequences for at least one man, Guillaume Edelin, the prior of Saint-Germain-en-Laye, who made the mistake of saying it was impossible to make pacts with the devil. He was arrested for covering up his own pact with Satan by saying such a ridiculous thing, forced to confess and even found to have handily written his own aforementioned pact. The document was "discovered" on his person. Realizing his best chance was to "repent", the prior was imprisoned for life rather than executed.

The "flying" was not, even then, always taken literally. Perhaps even more perverted than physically flying was the idea of preparing an evil "witches' ointment" to be smeared on the genitals (or other mucous membranes), inciting the sensation of floating, untethered from the body. Women, it was whispered, might even use it to lubricate a wooden dildo. Nothing, surely, could be worse than that?

The paste consisted of hallucinogenic plants known to be toxic: hemlock, nightshade, henbane and mandrake were regular ingredients. Worse, recipes often included fungus, which everyone knew was unclean: fly agaric was the usual suspect. Francis Bacon is said to have listed the fat of children "digged out of their graves" as the binding agent.

Spanish physician Andrés de Laguna (1499–1559) ran a test on one recipe in the early 1500s. He liberally smeared the local executioner's wife from head to foot with the concoction, then observed as she fell into a 36-hour sleep. Eventually, she was roused and was most unhappy to be woken. She claimed she had been in the middle of cuckolding her husband with a younger, lustier lover, surrounded by all the delights of the world. Terror by the god-fearing (especially male) world that it might have been infiltrated by (especially female) consorts of the devil, skyrocketed.

Opposite Woodcut by Ulrich Molitor from his 1489 book, *On Witches and Diviner Women*, showing three witches as animals flying on a forked branch.

128

Ag. muscarius

AMANITA

Fly agaric
Amanita muscaria

The ultimate fairy-tale toadstool, with its bright red, white-spotted cap, the fly agaric is the most recognizable – and most illustrated – of all fungi.

It is hardly surprising that its folkloric history runs deep and wide. *Amanita muscaria* has also been known for millennia as a hallucinogenic. Its various folk names across the world refer to three things: flies, madness and diabolic possession. It certainly looks suspicious, with that shiny red cap, white speckles and matching white gills and stipe – but, oh, what a beautiful devil. Commonly found across temperate and cooler parts of the northern hemisphere, it is associated with pine and birch forests and, being mycorrhizal, is difficult to cultivate, meaning it must be gathered.

Amanita muscaria labours under a persistent myth that it is a natural fly killer. In German, it is the *fliegenpilz* (meaning "fly mushroom"), while in parts of rural France it is the *tue-mouche* ("kill-fly"). The Russians call it *mukhomor* ("fly killer") and there are similar names in the Netherlands, Scandinavia and Eastern Europe. In Japan, *hayetoritake* means much the same thing. Its English name, fly agaric, seems to be a more recent addition to the dead-fly bandwagon. Perhaps the belief comes from the thirteenth-century German friar Albertus Magnus, whose collected works in science and philosophy run to 38 volumes. He twice refers to the insecticidal powers of *A. muscaria* – the odd thing being that it does not appear to be true. Despite this being pointed out in 1779 by Jean Baptiste Bulliard, who even called for a name change, no one took any notice.

Valentina Wasson tells of a Russian friend's childhood nurse, "an unlettered peasant woman", who nightly put out a saucer filled with crushed and sugared fly agaric. When the child pointed out that the flies were just eating the sugary mess, she would reply, "They are sure to die later." Darker and more mysterious are the legends referring to the mushroom's psychoactive qualities. It is poisonous, though only occasionally deadly, and humans have used its edgy toxicity for rituals and recreation for millennia.

The Koryak people of Siberia tell of Big Raven, who used the fly agaric sent to him by Existence to fly, enabling him to deliver a heavy bag of food so a whale could return to the sea with its family. Impressed, Big Raven caused the fungus to stay so his children could find visions for themselves. As a safety measure, a human shaman would ingest the mushrooms for the full experience, and then everyone else drank his urine.

Opposite Fly agaric (*Amanita muscaria*), Kew Collection.

The Koryaks were by no means the only people to use the fungus for its psychoactive properties; the Wassons discovered evidence in cultures across Scandinavia, Siberia and much further afield. In 1559, Franciscan missionary Bernardino de Sahagún described a Mesoamerican ceremony where partakers consumed mushrooms they called *teonanácatl*, or "flesh of god". Many links with ancient civilizations have been proposed, including references in early Taoist literature, the possibility of *Amanita muscaria* as ambrosia, the mythical nectar of the ancient Greek gods, and even one of many possible main ingredients of the early Hindu ritual drink, *soma*.

Humans are not the only creatures affected by it – reindeer actively seek out the fungus, and on ingestion behave as though drunk, twitching their heads and staggering erratically. Perhaps because of their bulkier constitutions, they do not suffer any long-term harm. It's even said that their meat has mildly hallucinogenic effects. Thanks to mammals' inability to break down the mushroom's active agent (muscimol), its effects remain potent in urine. It is thought that humans observed reindeer drinking each other's urine and adopted the practice for ritual use. The urine can be "recycled" several times. Do not try this at home.

Amanita muscaria has been closely associated with winter festivals, especially Christmas, for many years. It is a favourite image for Christmas cards, perhaps because it is often found growing beneath spruce (*Picea sp.*) trees, or that it is one of the few bright colours in an early winter forest. Such greetings are particularly popular in Scandinavia, where they usually also include *tomte* or *tonttu* – gnomes – either sitting on flycaps or using them as little houses.

In Germany, chocolate and marzipan *glückspilze* – lucky mushrooms – are exchanged as gifts on New Year's Eve, a modern version of the dried toadstools said to be worn as amulets a hundred years ago. One relatively recent theory about the origins of Father Christmas is that he represents the shamans of Arctic and Siberian regions, and that his red coat trimmed with white fur is reminiscent of fly agaric. It has been posited that his gifts of hallucinogenic mushrooms led to visions of flying reindeer.

Containing the psychoactive compound ibotenic acid and traces of the sedative-hypnotic muscimol, fly agaric is not all *tomte* and *glückspilze*. Ingestion results in reactions that run the gamut between euphoria and exhaustion, and can be very unpleasant. Any vivid dreams and sensations of weightlessness are cancelled out by less enjoyable symptoms, including incoherent speech, vomiting, diarrhoea, rapid breath, slowed pulse, dizziness, drowsiness, headaches, seizures and, possibly, coma or even death.

Opposite *Amanita* sp. from M. A. Burnett *Plantae Utiliores*, 1842–50

Amanita muscaria, var.

Ergot

Claviceps purpurea

*Before science figured out the role of pathogens in nature,
people turned to other explanations for strange occurrences,
disturbing diseases and peculiar behaviour.*

Few organisms have been blamed for as many frightening phenomena as ergot.

Claviceps is a genus of fungi that principally grows on cereal crops, especially rye (*Secale cereale*). It is an endophyte, living between the plant's cells, gaining nourishment from its host and, in return, protecting the grass from insect damage. When ingested, alkaloids present within the fungus, such as ergometrine, ergosine and ergotamine, can cause a serious reaction in both humans and animals.

Such symptoms have been described since ancient times, and variously interpreted, depending on their severity, as a good or (usually) bad thing. Scholars have found mention of similar-sounding reactions in texts from ancient China, Assyria and Egypt. The ancient Greek physician Hippocrates may have been describing ergot poisoning when he talked of the plague of *melanthion*, even as the fungus was also being used to stop bleeding after childbirth. Some have also suggested it was taken as a hallucinogen in religious rites. Traces of ergot discovered in prehistoric bodies preserved in bogs have led to suggestions that the fungus may have been used in ritual murders.

There are two main ranges of symptoms of ergot poisoning; both are devastating. Convulsive ergotism

sufferers experience nausea and vomiting, extreme joint stiffness, muscle spasms, itching, seizures and severe diarrhoea. These are often accompanied by hallucinations and psychosis. Gangrenous ergotism is where the blood vessels to the extremities are constricted to the point where fingers and toes lose sensation and even fall away without causing any pain, since the nerve has rotted away. A burning sensation in the disease's initial stages was known in medieval times as *ignis sacer*, "holy fire". It was thought that God was punishing sinners, who were now passing into hell before they had even died. Many did not have to wait long for death. Hieronymus Bosch (1450–1516) depicted sufferers of "St Anthony's Fire" writhing in pain, covered with ulcers and lesions and tormented by fantastical monsters.

Bosch was only one of many artists inspired by the agonies of St Anthony the Great, the third-century monk tempted by demons in the Eastern Desert of Egypt. Salvador Dalí's 1946 surrealist masterpiece *The Temptation of St Anthony* depicts a naked Anthony brandishing his cross at a terrifying menagerie of spindle-legged horrors, set against

Opposite Rye infected with *Claviceps purpurea* causing the growth of ergots, from P. Bulliard *Herbier de la France*, 1780-98.

LE SEIGLE COMMUN *FLO. FRAN.*

cale cereale *L. S. P. 124. cette plante est annuelle on la cultive partout; ses tiges s'elevent de 5 a 6 pieds ont*
epi terminale composé de 36 a 48 épillets qui ont chacun deux fleurs A. chaque fleur B a deux valves dont l'ex-
rieure est barbue, trois étamines et un germe surmonté de deux styles velus, on rencontre toujours deux valves calici-
illes aa. a la base de chaque épillet.

?B. La fig. M est celle d'un epi de SEIGLE. en fleur. la fig. N est celle d'un epi de SEIGLE. chargé de bonnes graines et D'ERGOTS. les fig.
3. representent deux epillets, une des deux fleurs a été retranchée de l'epillet B. la fig C est celle du germe. la fig D celle des graines. les fig. E. F.
t celles de toutes les formes D'ERGOT.

FRANCE on mange surtout dans les campagnes autant de pain de SEIGLE que de pain de FROMENT, il y a des
nées ou la maladie du seigle qu'on nomme ERGOT, CLOU, BLED CORNU, cause les accidents les plus facheux
y. Mem. Soc. Roy. de Med. par M. l'ab. TESSIER page 417 et les discours sur les plantes alimentaires et sur les
ntes vénéneuses de la FRANCE.

an innocent, pale-blue sky. Dalí painted it for a competition; the winning canvas would appear in the movie *The Private Affairs of Bel Ami*, an adaptation of a work by the French author Gustav Flaubert who, in turn, had been obsessed with the saint for most of his life. Max Ernst eventually won the contest; medieval skin-disease sufferers would have related to his vision, filled with horrific fiends, each "more Bosch than Bosch".

There were at least 130 ergot-related epidemics in Europe between 591 and 1789, killing and maiming tens of thousands at a time. An episode in France in the year 944–45 saw 20,000 casualties in Aquitaine and Paris. Outbreaks were particularly common just after the Black Death (1346–53) began to spread; indeed, ergotism shares many similar symptoms to plague. Some scholars have suggested that the Black Death would have had its worst effects in communities whose immune systems had already been compromised by ergot poisoning.

Cases nearly always occurred in rye-dependent countries such as Scandinavia, Germany, France and Eastern Europe. No incidences were recorded in England, where the crop has never been a popular one, but in certain parts of Russia, ergot was still killing people up to the Second World War. The worst years for the disease coincided with particularly cold winters followed by wet springs, when the fungus was at its most virulent. Rye was commonly eaten by those who could not afford white wheat (which was more rarely infected), so St Anthony's Fire was thought to be divine judgement on the poor, who could only manage grey or black flour. It seemed to affect young people in particular, possibly for the simple reason

Opposite Painting depicting witches flying menacingly over a town.

that they ate more of the poisoned bread.

The monastic Order of the Hospitallers of St Anthony was founded around 1100, in Grenoble, France, as thanks for the delivery of a nobleman's son from the disease via the holy relics of the saint. A series of "Antonine" hospitals became famous across Europe for tending sufferers of St Anthony's Fire, as well as leprosy and other skin diseases. The main treatment was prayer, though herbal healing ointments were also used, along with copious quantities of pork fat rubbed onto the affected limbs to ease the inflammation. The best rates of healing were often at hospices in areas where rye was not widely grown.

The cause of such symptoms was not initially traced back to infected rye. Indeed, for some communities, the blackened pegs, or sclerotia, infiltrating healthy ears of rye were thought to have magical or healing powers. In parts of Poland, people thought that large amounts of blue-black on the crop signalled a good harvest. Crop yields may well have been abundant, thanks to rain (providing the humidity ergot needs), but they would have been potentially deadly. In the 1740s, some religious people stopped suggesting that ergot poisoning was possession by demons and instead saw the hallucinations as a holy experience or even a religious ecstasy.

Folklore, however, felt there was something fishy going on in the rye. In Germanic tradition, the many *feldgeister* ("field spirits") that attack various crops include the *roggenwölf* ("rye wolf") and, even more terrifyingly, its mistress/alter ego, the *roggenmuhme* or "rye aunt". This female demon, also known by dozens of similar names, has fire in her fingers and tar in her pendulous, iron-nippled breasts, which she flings over her shoulders as she attacks the crop. She will also steal children who collect cornflowers among the rye. Among the many counteractive precautions that vary according to region, farmers would trap her spirit inside the last sheaf at the end of the harvest. This was sometimes ritually "beheaded" or paraded through the village. In some communities, it was burned the following spring.

German physician Wendelin Thelius first worked out the relationship between harvests and ergotism in 1596, while discussing an outbreak in the Duchy of Hesse. In 1676, French botanist Denis Dodart confirmed the culprit was infected rye, but it took until 1853 for French mycologist Louis Tulasne to prove Augustin de Candolle's 1816 theory that the infection was fungal.

In 1917, Swiss biochemist Arthur Stoll isolated the various alkaloids from ergot for medicinally therapeutic purposes, including heart disease and migraines. He also showed how some old folk cures, such as staunching blood after childbirth, actually worked. Stoll worked with fellow chemist Albert Hofmann, who was more interested in a substance they called lysergic acid, intending to find a stimulant that might work for circulatory and respiratory complaints. Instead, on 16 April 1943, he discovered LSD.

The curious phenomenon known as the Dancing Plague has often been attributed to the hallucinogenic effects of ergot poisoning. Occurring periodically throughout Europe from around the seventh to the seventeenth century, when it suddenly ceased, St John's Dance, or St Vitus's Dance, saw a kind of collective mania where people jerked and danced erratically and compulsively for hours, days and, on occasion, months, until they either recovered, collapsed or died.

This strange mass hysteria was characterized by groups "dancing" between villages, and often

Opposite Study of ergotism symptoms, after Theodor Otto Heusinger, 1856.

occurred in times of want. It was said sufferers begged clergymen to save them from the invisible devils that were tormenting them; for many, this was neither a joyous nor an ecstatic experience. Sometimes musicians were brought in to try to cure them but often had the opposite effect, as even more people joined the dance. It was particularly prevalent across Northern Europe, and may have provided inspiration for the folk tale *The Pied Piper of Hamelin*.

In Italy, a similar condition was known as tarantism, in which victims were believed to have been bitten by a tarantula, inspiring the tarantella folk dance. Sufferers were said to hate the colour black. Ergot poisoning is only one of the possible modern explanations for this phenomenon. Other scholars have suggested it was a tremor-inducing disease, such as epilepsy or encephalitis; that it was a kind of mass therapeutic release from hunger and poverty; that it was a ritualistic exercise; or even that it was staged.

In 1976, American psychologist Linnda Caporael suggested that the notorious Salem Witch Trials were the result of ergot poisoning. Rye was the staple crop of the fiercely religious New England community. Weather records tell us it would have been harvested and stored under damp conditions for some months when, in February 1692, eight young women accused certain neighbours of witchcraft. This would have been plenty of time for the fungus to have got a hold, affecting humans and animals alike. Caporael argues that both accusers and accused may have been acting under the effects of the poisoned rye, causing symptoms such as hallucinations, vertigo, crawling sensations on the skin and muscle convulsions. The following summer was hotter and drier, and the "witchcraft" ended, but it was not enough to save fourteen women, five men and two dogs from execution.

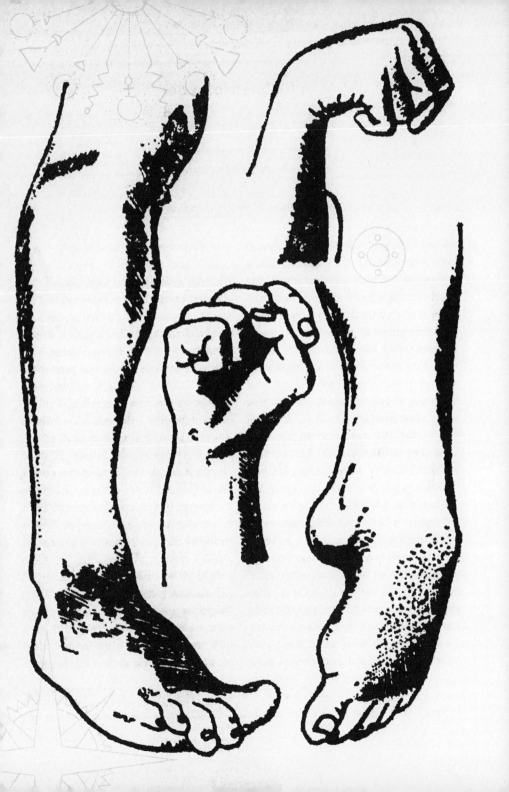

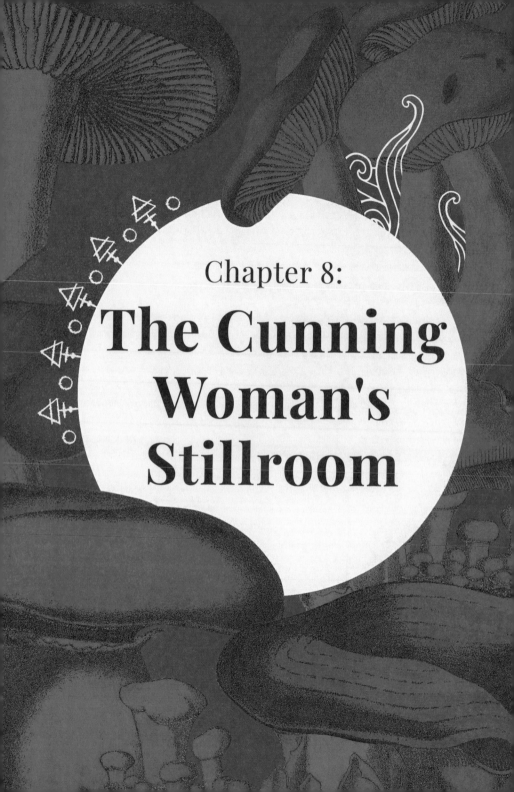

Chapter 8:
The Cunning Woman's Stillroom

The classic stillrooms of previous centuries, where the mistress of the house would concoct tinctures, tisanes, treatments and tonics for the whole family's health did not use fungus as much as herbs and plants for one main reason: in many countries, mushrooms were not considered trustworthy. This did not stop some, however, and traditional and folkloric medicine has found many uses for this fascinating kingdom, from good-luck charms to love philtres.

The best form of medicine has always been to avoid illness in the first place. Preventing disease was by far the most useful thing a fungus could do.

In Korea, the juice from a cooked mushroom has long been considered an excellent all-purpose elixir, while in several parts of the world, tonic made from regular field mushrooms (*Agaricus campestris*) was thought to make a general energizing health drink that would prevent tuberculosis and colds. Boiled in milk, it soothed sore throats. Jelly fungus, such as witches' butter (*Exidia glandulosa*) is considered useful in TCM to improve circulation.

Sometimes just wearing or displaying a fungus has the effect of a good-luck charm. Wearing cramp balls (*Daldinia concentrica*, aka King Alfred's cakes) under the armpits has been associated with eliminating cramps since ancient times. Dried, the strange black spheres, which are found on ash trees, also made useful tinder. Archaeologists have discovered amulets in the shape of mushrooms in several cultures, including a little stone mushroom from Turkey, from around the fourth to third century BCE, perhaps worn for protective or health purposes, as well as decorative ones. The famous mushroom of immortality, the lingzhi or reishi (*Ganoderma lucidum*) was sometimes

Agaricus campestris, *Linn.*

hung above entrances as a talisman against evil spirits. Even its image was protective. In Zoroastrianism, the reishi shape was used as a protective motif in clothing, including children's smocks.

Robert Rogers's extraordinary 600-page manual *The Fungal Pharmacy* (2011) is testament to the sheer number of traditional medicines found across the world. Many of them have been in use for centuries, while more modern applications have been found in practices such as homeopathy. Each species may be used for several purposes. For example, *Amanita muscaria* is, for the First Nation Cree, most useful as the ingredient in an eye wash, while in European traditions, it may be a gargle for sore throats. Jelly ear (*Auricularia auricula-judae*) has been used for much the same purpose and, across Europe, boiled in milk to relieve jaundice.

It is just as well to ignore either a fungus's name or appearance when considering its potential medicinal properties. *Pisolithus arhizus*, the dyemaker's puffball, also goes by the unedifying names of dead man's foot and dog turd fungus, yet it has been used to staunch anything from watery chilblains to stomach bleeding.

Many fungi are said to hold antibacterial qualities. The most famous of these is, of course, *Penicillium*, the mould from which the world's first antibiotic, penicillin, is derived. Also numbered among such fungi are St George's mushroom, *Calocybe gambosa*, which some studies have shown to have antibacterial properties. Charcoal made from the birch polypore (*Fomitopsis betulina*) was used as an antiseptic in Sussex. This would have been particularly useful if the blades sharpened on the dried fungus (another of its names is the razor strop

Opposite Common field mushroom (*Agaricus campestris*) from Anna Maria Hussey *Illustrations of British Mycology*, 1847–55.

fungus) turned out not to be quite cutthroat enough.

The race for cancer treatments has focused a spotlight on many fungi species, and virtually all the main families have been researched to explore their possibilities. Promising species include *Grifola frondosa* (hen of the woods), *Phellinus igniarius* (bracket fungus), *Hericium erinaceus* (lion's mane) and *Hypsizygus tessellatus* (white beech mushroom). While research is often in its early stages, there are several traditional medicines that lay claim to antitumor and antioxidant properties. Shiitake (*Lentinula edodes*) has been used in Japan for the treatment of gastric cancers for some time. Cyclosporin was first isolated from the beetle-parasitizing micro-fungus *Tolypocladium inflatum* and used in treatments for rheumatoid arthritis, Crohn's disease and as an anti-rejection drug after organ transplants. It features on the World Health Organization's List of Essential Medicines.

Many fungi look strangely like human reproductive organs, so it is unsurprising that several have been considered aphrodisiacs. The Haida people live on an archipelago off the coast of British Columbia, Canada. According to Haida folklore, Fungus Man was born from a bracket fungus on which the Creator, Raven, had painted a face. Mycologist Lawrence Millman tells how only Fungus Man was able to break through spiritual barriers to find the genitals of women. Tibetan physician and botanist Zurkhar Nyamnyi Dorje discussed the aphrodisiacal qualities of *yartsa gunbu* (*Ophiocordyceps sinensis*), the fungus that grows out of the heads of dead caterpillars, as far back as the fifteenth century. It is now a multimillion dollar industry (see page 146).

Across Asia, *Ganoderma lucidum* (lingzhi) was given to men by women who were interested in them sexually; Robert Rogers tells us this was often via a go-between. In Lapland, youths were a little more direct. Carl Linnaeus talks, in 1737, of how a young man would preserve a piece of *Trametes odora* (*Haploporus odorus*) "in a little pocket hanging in front of his pubes" in the hope that its spicy, pungent smell would attract the girl of his dreams. Linnaeus reflects that in other countries girls were wooed with "coffee and chocolate, preserves and sweetmeats, wines and dainties, jewels and pearles, gold and silver, silks and cosmetics, balls and assemblies, music and theatrical exhibition," while in Lapland "you are satisfied with a little withered fungus".

Once the loving couple had got together, fungi had a part to play in conceiving a child. *Leechdoms, Wortcunning and Starcraft of Early England* is, despite its name, a collection of magical recipes made by a nineteenth-century vicar, the Reverend Oswald Cockayne, subtitled *The History of Science in the Country Before the Norman Conquest*. It is hard to know the exact provenance of the documents Cockayne published in 1864, but the recipes are intriguing. "To make a woman pregnant," he writes, "give to drink in wine a hare's runnet (stomach lining) by weight of four pennies, to the woman from a female hare, to the man from a male hare, and then let them do their concubitus, and after that let them forebear; then quickly she will be pregnant and for meat she shall some while use mushrooms, and instead of a bath smearings; wonderfully she will be pregnant." Alas, instructions on which mushrooms the woman was to consume – or what "smearings" might have consisted of – are lost to antiquity.

Opposite Hen of the woods (*Grifola frondosa*) from Anna Maria Hussey *Illustrations of British Mycology*, 1847–55.

Below St George's mushroom (*Calocybe gambosa*) from Anna Maria Hussey *Illustrations of British Mycology*, 1847–55.

Polyporus intybaceus, Fries

Caterpillar fungus

Ophiocordyceps sinensis

It is said that in ancient times, Tibetan herdsmen noticed that rutting season began soon after their animals' hooves dug up soil containing a strange fungus.

Figuring that what worked for lady yaks might work for humans too, they began to harvest *yartsa gunbu* as an aphrodisiac.

Today, caterpillar fungus is a multimillion dollar business and the most important cash crop in certain rural areas of the Tibetan plateau. In Japan, it is *tochukaso*; Chinese people know it as *dōng chóng xià cǎo*, meaning "winter worm, summer grass", because it was once thought to be an animal in winter and a plant in summer.

Ophiocordyceps sinensis (once *Cordyceps*) is part of the *Ophiocordyceps* genus, which parasitize insects (see also zombie ants, page 188), usually, though not always, killing their host. *O. sinensis's* prey is usually the larvae of the ghost moth (*Hepialidae*), which live underground, feeding on roots. In late summer, they shed their skins, leaving them vulnerable to fungal spores in the soil. *O. sinensis* attacks the larva and slowly feeds off it, using a network of hyphae to liquidize then digest the internal organs, effectively mummifying the creature. The following spring, a long, dark brown stroma (fruiting body) grows out of the corpse's head, ready to disperse its spores.

O. sinensis is said to have been a favourite of the mythical Yellow Emperor of China. It appears in Tang dynasty texts from 620 CE and is mentioned in Ang Wang's 1694 *Ben Cao Bei Yao (Essentials of the Materia Medica)*. *Ophiocordyceps sinensis* is still used in both TCM and Tibetan traditional medicine for many conditions, as well as still being considered an aphrodisiac. It attracts very high prices – according to the *Journal of Ethnopharmacology*, high-quality specimens cost $140,000 per pound in 2019. Although a great incentive for indigenous communities to harvest as much as possible, this has also led to political turf wars, and caterpillar fungus is now an endangered species. Since wild specimens can carry high levels of arsenic and other heavy metals, caterpillar mushrooms are not necessarily good for humans either. A close relative, *Cordyceps militaris*, is now cultivated as a cheaper alternative, even if its active ingredients are somewhat different. As with so many fungi, Western science is now taking a great interest in the entire *Cordyceps* genus for its anti-inflammatory qualities and transplant recovery. Ghost moth larvae are set to have headaches for some time yet.

Opposite Caterpillar fungus (*Ophiocoryceps sinensis*) by Katie Scott from *Fungarium*, 2019.

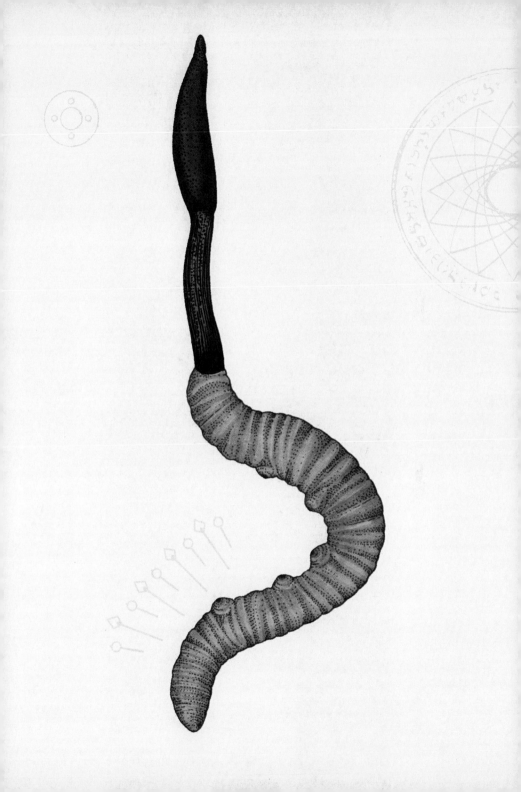

Exidia Auricula - Judæ. *Linnæus.*

Jelly ear
Auricularia auricula-judae

An uncanny growth habit of Auricularia auricula-judae *means that wherever this curious fungus appeared in the world, it has been associated with the human ear.*

Seeming to sprout directly from old wood, it's variously known as wood ear, jelly ear, black wood ear and pig's ear, but its most common folk associations are with the biblical traitor Judas Iscariot.

The fungus was traditionally – and erroneously – thought to grow only from the "unlucky" elder (*Sambucus*), the tree from which Judas hanged himself after betraying Christ. Anyone claiming that the small and relatively feeble elder is unable to bear the dead weight of a man would be told that God punished the once-handsome tree by shrinking it, and shrivelling its giant berries to tiny beads. The name "Judas ear" is consistent throughout Europe – in German, it is known as *Judasohr*, in French *oreille de Judas*; it is claimed the less-savoury English alternative "Jew's ear" comes from a medieval mistranslation of "Judas".

Scientific nomenclature has reflected the fungus's uncanny likeness to an ear, too – pale brown, slightly translucent, cupped and velvety on the outside – though it has taken a while to get to the modern name. Carl Linnaeus bunched together many gelatinous entities as *Tremella* – "to tremble". The group became unwieldy and imprecise, burgeoning to include jelly fungi, slime mould and seaweed. Eighteenth-century French mycologist Jean Baptiste Bulliard included the Judas connection in his re-jig, *Tremella auricula-judae*, but it took until 1888

for German scientist Joseph Schröter to separate jelly ear into a smaller, more specific genus, *Auricularia*.

A. auricula-judae grows on elder and other, usually deciduous, hardwoods and some conifers such as spruce (*Picea*). Medieval physicians found the fungus interesting in their doctrine of signatures. Jelly ear, with its gelatinous constitution, was soaked overnight in rosewater and used as a remedy for eye problems. It was also thought to have a throat-like inner, so Elizabethan herbalist John Gerard suggested boiling it in milk to soothe sore throats.

A close relative, *Auricularia nigricans*, is known in China as cloud ear, perhaps for its similarity to brush-painted clouds in art or its ability to swell in water. In Thailand, it is rat's ear; in Japanese, *arage kikurage* roughly translates as "tree jellyfish" or "hairy forest jellyfish". It is often sold dried, in large bags unimaginatively labelled "black fungus" but although the taste is bland, rehydrated it brings texture to a number of Asian dishes. Also used in TCM, the fungus is said to carry many health-giving properties, including high levels of iron and vitamin K, the ability to lower blood pressure, and boost mental and physical energy.

Opposite Jelly ear (*Auricularia auricula-judae*) from Anna Maria Hussey *Illustrations of British Mycology*, 1847–55.

Amethyst deceiver

Laccaria amethystina

Looking like a purple velvet toadstool from a poster on a 1960s student bedroom wall, the amethyst deceiver is one of the most beautiful fungi on the woodland floor.

Sometimes. Look again, and this curious mushroom lives up to its name: deceiver.

Fungi with the genus name *Laccaria*, from the Persian word for "painted", are known in many languages as "deceivers" for their chameleon-like capacity to change their appearance as they mature. All fungi alter with age, but these can even look different from each other at the same stage in their lives. This makes it challenging for the beginner forager to be sure of what they have found. Deceivers are ectomycorrhizal; i.e. they have a symbiotic relationship with plants – often beech or oak – but do not penetrate their host's cell walls. Instead, they form a network or "hyphal sheath" around the roots, allowing the tree to more easily ingest water and minerals.

The amethyst deceiver (*Laccaria amethystina*) can be anything from lilac to deep purpleits cap may be convex or concave. *Laccaria bicolor* often appears in classic "mushroom" shape, with a flattened cap and fibrous stipe of medium thickness. It can have a soft, velvety appearance and may be anything from reddish mauve to the regal violet of a Roman senator's toga. Then again, it may curl up at the edges, its wide-spaced, long-short purple gills curling into a frill. As it matures, especially after hot days, it begins to dry out, fading to a dull brown or pale buff colour. Amethyst deceiver is edible but not considered flavourful. Indeed, some believe it to be a bioaccumulator of arsenic, so, as Robert Rogers suggests, "There are lots of better-tasting and safer mushrooms."

One of the best reasons to avoid gathering it is that, due to its shapeshifting abilities, a mature amethyst deceiver can sometimes be be confused with the purple form of the lilac fibrecap (*Inocybe geophylla*), with potentially fatal results. The violet webcap (*Cortinarius violaceus*) is also purple, sometimes so deep in hue it is almost blue-black. A real "fairy toadstool", it is a much chunkier mushroom, rounder and with a thicker stem. It is related to poisonous species, however, and very rare. If picked in error, the loss is even worse for the ecosystem.

Amethyst deceiver sometimes grows alongside *Laccaria laccata*, known simply as "the deceiver". Less flashy but still a fan of fungal fancy dress, it (usually) starts out orange-brown, fading to pinkish beige. The twisted deceiver (*Laccaria tortilis*) is also orange-brown with a depressed cap that matures to a wavy, coral-like frill.

Opposite Amethyst deceiver (*Laccaria amethystina*) from James Sowerby *Coloured Figures of English Fungi or Mushrooms*, 1795–1815.

Witches' butter

Tremella mesenterica

Confusingly, this group of similar-yet-not fungi is known by several common and Latin names.

The most obvious fungal candidate for witches' butter is *Tremella mesenterica*, also known as yellow brain and star jelly. Bright yellow in colour and appearing to drip from its hardwood host, for centuries this gelatinous fungus was thought to be the by-product of falling stars. The poet John Donne refers to it in an epithalamion – a poem written to celebrate a marriage – when he writes of the bridegroom running for luck:

As he that sees a starre fall, runs apace,
And findes a gellie in the place

An even more common idea was that it was butter made by fairies or elves. In Sweden, it was the work of trolls who milked the farmers' cows and splashed the butter while churning. On Midsummer Eve (23 June), it was traditional to throw the fungus into the fire, forcing the troll to appear and beg for mercy.

The nineteenth-century folklorist Thomas Keightley describes a Finnish tradition whereby the fungus is fried with tar, salt and sulphur before beating it to make the witch's kobold (a goblin-like creature who had excreted or vomited the fungus) appear. The witch would then arrive, too, begging for its life.

The seventeenth-century Welsh Courts of Great Sessions chronicle several "witchcraft as malefice" cases. Part of the charges against Gwenllian David, accused of witchcraft in 1656, included her alleged use of witches' butter for evil purposes. To prove her guilt, a red-hot knife was thrust into the fungus on the doorpost of her house. The elderly woman was apparently in such instant pain that, after two weeks, villagers removed the knife out of pity and "immediately the said Gwenllian began to recover and further this examinant deposeth not".

The slime mould *Fuligo septica*, rejoices in the common names dog vomit, slime mould and scrambled egg slime and, in Estonia, witch's shit. A protozoa, not a fungus, it is not related to the "butters", whose colours include brown witches' butter (*Phaeotremella foliacea*), black witches' butter (*Exidia glandulosa*), orange jelly (*Dacrymyces chrysospermus*) and apricot jelly (*Guepinia helvelloides*), each with its own properties and uses.

In China, *Naematelia aurantialba* is harvested from a liquid culture and used to add texture to walnut cakes, biscuits, noodles and bread. The silver ear, white jelly or snow fungus *Tremella fuciformis* is the only mushroom considered suitable to be sweetened as a dessert, enjoyed in China, Japan and Korea.

Opposite Witches' butter (*Tremella mesenterica*) from Anna Maria Hussey *Illustrations of British Mycology*, 1847–55.

Fungus in the garden – the white hats

Digging is a way of life for gardeners. G.T. McKenna, in Allotmenteering in Wartime *(1944), advises his reader to mark out a space 2 feet wide by 15 feet long, dig out the soil to at least the depth of a spade, breaking up the earth.*

They are to do the same on the other side – move all the earth from the second trench into the first. Working their way along the entire plot, they replace 45 trenches on each side with the other's soil. The last trench is replaced with the contents of the first.

Apart from being backbreaking work, this centuries-old, standard advice for gardening is now thought to do more harm than good. Some cultures, such as Russia, have always known the relationship between fungus and trees, often calling a mushroom by the name of the tree it is associated with. Alas, until recently, no one has known the reasons for that relationship, or that it extends to about 90 per cent of the plant kingdom.

Mycorrhizal fungi live in almost every inch of the soil below our feet, colonizing plant roots. In return for a home and carbohydrates photosynthesized by the plant, the fungi form networks of hyphae around the roots, supplying the plant with water and vital nutrients including phosphorous, magnesium and calcium. Different plants attract different kinds of mycorrhiza, often depending on the acidity of the soil. Mycorrhizas bond with around 90 per cent of plants. Around two per cent, mainly tree species, which dominate many land-based ecosystems, actually form relationships with ectomycorrhizal fungi. These often form the world's most popular edible mushrooms, such as truffles, chanterelles and porcini, one of the reasons why they are hard to grow as a commercial crop.

Mycorrhizal fungi are largely invisible to the human eye, though when they do appear, often as a white, stringy mycelium in the soil, they have been assumed to be a bad thing and broken up. Alas, humankind's millennia of digging the soil has seemingly been counterproductive and our best policy may be to leave the soil to its own devices. If we must dig or plough, we should steer clear of mechanical diggers, which pulverize fungi. Leaf mould piles benefit from similar neglect. Unlike compost, which relies on bacteria to break down plant waste into usable soil, leaf mould is exactly that – a mould – breaking down tough leaf matter via fungal decay, creating a friable (easily crumbled) soil conditioner with many uses.

The no-dig philosophy, at first pooh-poohed as being a "lazy person's" theory, is gaining traction as we realize how important it is to promote the underground relationships between plants and fungi, about which we still have much to learn.

Opposite Golden chanterelle (*Cantharellus cibarius*) by Elsie M. Wakefield, c.1915–45.

Fungus in the garden – the black hats

While gardeners need certain fungi to make their crops flourish, the number of destructive species always seems to outnumber them.

Since the dawn of time, farmers have battled blight, black spot, crown-rot, mildew and many others, often with only folk cures as weapons. Few food crops are resistant to attack.

Scientists now believe the catastrophic ancient Egyptian crop failures, leading to the mass migration of Israelites described in the Bible's Book of Exodus, may have been due to parasitic fungal rusts. A wide-ranging group of species, there seems to be a rust for every plant from leeks to pears, soybeans to wheat. Powdery mildews are another gardening horror. Members of this group block the stomata in the green parts of plants, reducing their photosynthetic ability, appearing as chalky patches of whitish hyphae on leaf surfaces.

Folklorist Susan Drury tells us that in Hertfordshire a branch of blackthorn (*Prunus spinosa*) used to be cut before dawn. Part was burned in the fields and part hung in the house, either as a charm against mildew or ready to be burned if signs of the disease appeared. Much of the problem lies in watering; if the leaves, rather than the ground, become wet, the fungus has more chance to grow, especially in warm, damp conditions. Sometimes old gardeners' lore hits the mark, such as the adage that if it rains with sunshine, black spot is sure to follow. Trees are by no means immune to fungal pathogens. Dutch elm disease was originally thought to be caused by members of the *Scolytus* family of elm bark beetles. While the real culprit is a fungus, *Ophiostoma novo-ulmi*, it needs its insect friend to carry it. Millions of elms have been lost, including nearly the entire UK elm population. While a new generation of saplings have sprung back, when they reach maturity, the beetle and its deadly stowaway will return.

Phytophthora infestans is at the root of one of the most notorious plant diseases in history. It is no longer classified as a fungus, having recently been moved to the Kingdom Chromista (phylum Oomycota), but for years it has been known to attack members of the *Solanaceae* family. It turns potatoes to a black, stinking mush, and shrivels tomatoes. This is annoying in a regular crop, but usually people have alternative foods to turn to. Between 1845 and 1852, the poor of Ireland had no such luxury. The crop failed again and again and the people starved, perished and emigrated in their millions. Little or no folklore remains of the Great Famine. Too painful to recall, it was referred to as "the time of the great wind", recalling an unrelated hurricane that blew just before the real horrors began.

Opposite *Phlebia radiata* (wrinkled crust), causes white-rot in trees. From Anna Maria Hussey *Illustrations of British Mycology*, 1847–55.

Honey fungus
Armillaria mellea

Bane of gardener and forester, the parasitic members of the honey fungus genus have much to answer for, but delving a little deeper reveals some ancient secrets and the largest organisms on the planet.

Armillaria comes from the word "bracelet", referring to a partial veil of tissue that covers the gills of most young honey fungus, protecting the spores. *Mellea* means "honey". The genus includes around 12 fungi, only about half of which merit their reputation as wood rotters – but for horticulturalists, that is enough.

Some *Armillaria* are saprotrophs, meaning they feed on dead matter. Others are parasitic, killing their host by attacking living cells then consuming the woody "lignin" part of the dead cells and leaving only soft, white cellulose as "white rot". They wait for their moment, then enter, perhaps through a lesion from poor pruning, insect or animal damage, or simply cracks in the wood after a dry summer. Honey fungus will attack hardwood, like oak and beech, as well as fruit trees, hedging and ornamentals, even vegetables if it gets a chance. Once attacked, even healthy trees have little chance of survival.

The first signs of infection may be small colonies of yellow fruiting bodies, ranging from little pixie hats to mushrooms 15cm across. The species vary, from the rich brown of the dark honey fungus (*Armillaria ostoyae*) and the pink-brown gills of the ringless honey fungus (*Armillaria tabescens*) to the golden caps and white stipes of *A. mellea*. The writing is on the bark. The tree will collapse and the honey fungus will move on. Occasionally, a plant may die for no obvious reason; if there are black bootlace-like rhizomorphs under the soil, it's honey fungus. *A. mellea* can travel great distances and it is thanks to these networks that the genus has gained its place in the record books as the largest living thing on Earth. One "humongous fungus" in Michigan, an *Armillaria gallica*, is said to cover 37 acres, weigh at least 400,000kg and probably date back 2,500 years. Yet this is a spring chicken in comparison to an *Armillaria ostoyae* in Oregon's Malheur National Forest, perhaps 8,000 years old and covering 3.7 square miles (2,368 acres). Another *A. ostoyae* measuring 2.5 square miles (1,600 acres) has been found in Washington state.

Armillaria mellea is edible and particularly popular in Eastern Europe. It must be thoroughly cooked as some find it hard to digest, perhaps lending the fungus its German name *hallimasch*, a contraction of *holle im arsch* or "hell in the ass".

Opposite Honey fungus (*Armillaria mellea*) by Florence H. Woodward, Kew Collection, late-nineteenth century.

17-9-86

Agaricus melleus.

Noble rot

Botrytis cinerea

Botrytis comes from the Greek words for "wine" and "disease" and
Botrytis cinerea is usually considered a villain.

The mass of grey fur on a punnet of strawberries, the hard, brown ball that was once a rosebud, the sickly shine of tomatoes blighted beyond redemption. There are, however, rare cases in which the very same mould is considered highly desirable.

Noble rot occurs when ripe grapes become infested with *Botrytis cinerea* under specific conditions. The fungus sucks the grapes' natural fluids, leaving a desiccated concentrate of sugars, acids and minerals. Although reduced, the harvest becomes intense, perfumed and very sweet.

Special weather conditions are needed to turn *Botrytis cinerea* from fiend to friend, limiting the regions where such wine can be produced. *Botrytis* needs rain to attack a crop, which can also only happen after the grapes are large enough for insects to puncture the skins for entry. If the climate remains humid, the fungus will get too much of a hold and the grapes will fuse into a grey mass called "bunch rot". In, often drier, conditions of late-autumnal Northern and Eastern Europe, however, the mould's progress is arrested enough for the grapes to shrivel, not rot.

The first written mention of the infection is in *Nomenklatura*, by Hungarian priest Balázs Szikszai-

Opposite Coloured woodcut of grape-gathering and wine-tasting, Johann Sittich, 1515.

Fabricius, but the legendary ancestor of Hungary's world-famous *Tokaji* was made in 1630. On learning the Turkish Ottomans were to invade her lands, Zsuzsanna Lórántffy, wife of Prince György Rákóczi, postponed her wine harvest. Left on the vine, the grapes became raisins. Her wine-maker Lázló Szepsi did what he could, soaking the shrivelled mass before fermentation. As is usual with folk tales, the story is much disputed, but golden and complex *Tokaji aszú* became the "king of wines and wine of kings".

Today, *Tokaji* is revered as one of the nation's most important *Hungarikums* or "cultural treasures". Aszú berries (botrytized grapes) are collected in *puttonyos* (large punnets). The wine is graded for sweetness by the number of puttonyos added to each cask. The rarest, *eszencia*, is made entirely from aszú berries. In Germany, their origin story begins in 1718. The grape pickers of Schloss Johannisberg couldn't harvest until they had permission from the Bishop of Fulda. The messenger bringing the order was attacked, and arrived after the grapes had rotted. The crop was written off and given to some peasants, who made a light yet surprisingly complex wine. The method of holding off the harvest has its risks – if it rains, the crop will be ruined – but in the years it is produced, *spätlese* (late harvest) is much sought after.

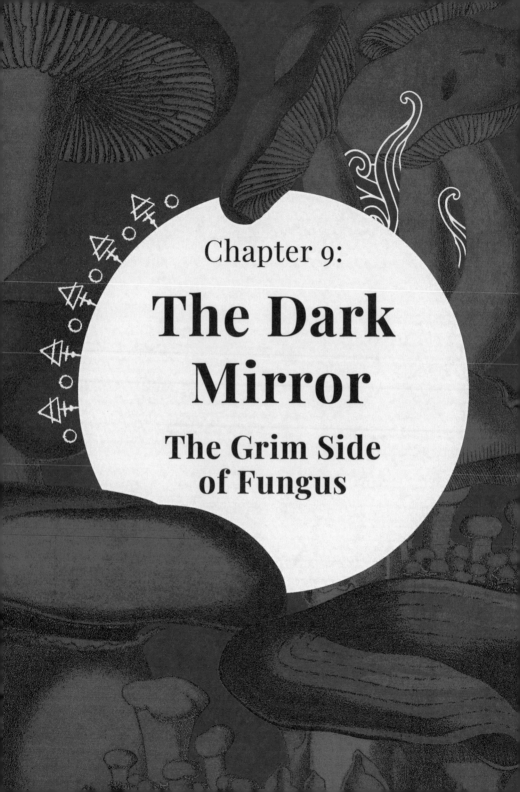

Chapter 9:

The Dark Mirror

The Grim Side of Fungus

Gordon and Valentina Wasson suggested that even the folklore of mycophilic countries is consistently more positive about mushrooms than in mycophobic cultures. For fungus-loving countries, fairy stories depict mushrooms sustaining lost wanderers in the woods or giving them magical powers. Fungus-fearful communities have traditionally seen "toadstools" as agents of witchcraft, poison and death. Even in the twenty-first century, for many, the darkness remains.

Much of the folklore in both mycophilic and mycophobic cultures stems from their respective attitudes towards education.

In mycophilic countries, children have traditionally been brought up to know which mushrooms are and aren't edible. Mycophobic communities prefer not to take the chance, and just tell their offspring to avoid all "toadstools".

In Northern European and American traditions, the poisonous aspect of fungi has held sway for centuries. Famous poisonings began in antiquity but instances continued, sporadically, throughout history. The death of the German Emperor Charles VI in October 1740 presents a classic example of possible *Amanita phalloides* (death cap) demise. The emperor suffered indigestion after eating a dish of mushrooms. He recovered, but 10 days later, took a turn for the worse and expired suddenly. Charles's death saw his empire descend into war; the writer Voltaire noted that the history of Europe had been changed by a dish of mushrooms.

According to Alisha Rankin's *The Poison Trials* (2020), Pope Clement VII authorized controlled tests of a new "medicinal oil" on condemned prisoners. Surgeon Gregorio Caravita gave the men marzipan cakes poisoned with wolf's bane (*Aconitum*, rather than a fungus). He then administered his new "miracle cure" to half the convicts, who largely survived to "live" as galley slaves. The rest died in horrible agony. Ironically, when Clement himself died, on 25 September 1534, some said he also succumbed to poisoning, via death cap mushroom. In truth, Clement probably just died from a long illness.

Perhaps part of the reason why so many people are scared of fungi is the number of fungal species that thrive on death. Performing the vital job of decomposing detritus that would otherwise smother the rest of life has given mushrooms a reputation as nature's undertakers. The fledgling forensic discipline of using fungus as trace evidence for establishing a victim's time of death may even provide a (so far, not very accurate) way to locate woodland graves and crime scenes.

Perhaps the darkest side of mycophobia manifested itself in 1938, with the publication of one of the most odious books of all time: *Der Giftpilz (The Poisonous Mushroom)* by Ernst Hiemer, published as Nazi propaganda. The children's storybook, equating Jewish people with toadstools, depicted wide-eyed "Aryan" children being duped in the forest by nice-looking but deadly mushrooms. Even solitary examples of these evil fungi could, the book warned, poison a whole nation. The accompanying illustrations of this appalling volume are in sickly sweet fairy-tale style and are not reproduced here.

Although individual poisonings have been known since antiquity, there has been a recent flush of academic papers warning about the possibility of the use of mycotoxins as a biological weapon. Although the ease of converting fungal spores to weapons is less than, for example, bacteriological pathogens, the matter is still being seriously researched, a sadly necessary precaution, some say, against potential terrorist activity.

Opposite Engraving of an octopus stinkhorn mushroom (*Clathrus archeri*).

Death cap

Amanita phalloides

The ancient Romans often enjoyed their mushrooms in "egg" form, before the immature fungus had a chance to ripen into individual adult form.

Many fungi, covered with a veil of anonymous-looking tissue, appear much the same when young, and foraging could therefore become a round of fungus roulette. The gatherer might be collecting, for example, some tasty cocorra (*Amanita calyptroderma*) – or something altogether less appealing.

There are numerous examples of death by *Amanita phalloides*. The Roman Emperor Claudius, in 54 CE, for example (see page 16), or the Holy Roman Emperor Charles VI of Austria (see page 164). Even when death cap was not being used deliberately as a murder weapon it could be an accidental killer for much the same reason – it looks very much like a number of tasty edibles.

Amanita phalloides now grows on every continent except Antarctica, having been spread by humans, possibly when importing wood or trees. The mycorrhizal death cap is mostly to be found under trees, especially oak and beech, though since tree roots can radiate a long way, the mushrooms are often found some distance from the tree's trunk, at the edge of the canopy. *A. phalloides* comes gift-wrapped, presented in the volva of white tissue around the base. Halfway up its slightly scaly white stem, the mushroom has a little skirt-like ring of membrane before opening out into a wide, convex cap with a sickly green metallic sheen and white gills. Cut open, death caps have a firm, pure white meatiness, which looks ideal to cook up for a meal. This is a bad idea.

Symptoms do not begin until some hours after ingestion, so the cause of the sudden illness is not always fathomed out in time to do much. Nausea, abdominal pain, vomiting, jaundice, seizures, diarrhoea and eventually coma may not begin for up to 40 hours. Sometimes, after being very ill, the patient appears to rally, only to suffer catastrophic organ failure soon afterwards. Dialysis and occasionally kidney or liver transplants may be necessary, but even this is often not enough to save the patient.

There is no way of making a death cap edible. Peeling the skin, boiling it to a mush or putting a silver spoon in the cooking water will not save a trip to A&E. First-century author Pliny the Elder suggested pears were an antidote to mushroom poisoning, and alcoholic perry drinks were popularly drunk after Roman feasts just in case someone's enemy had tried any funny stuff. Don't try this one at home. If you've accidentally eaten even part of a death cap, seek urgent medical help.

Opposite Death cap (*Amanita phalloides*) from JL Leveille *Iconographie des Champignons de Paulet*, 1855.

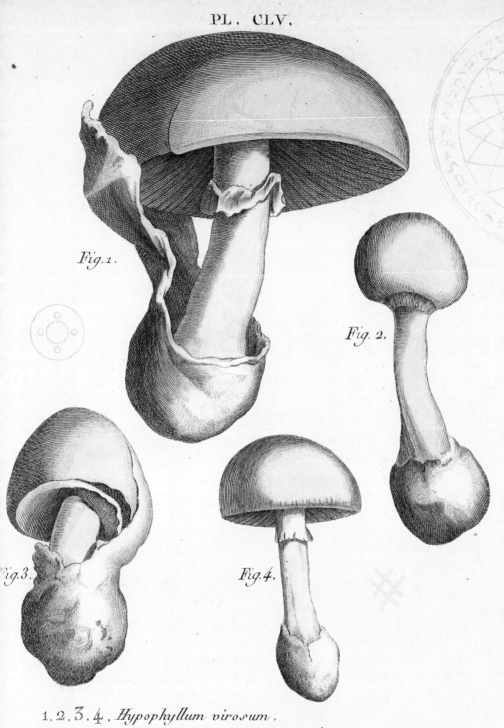

Fig.1.

Fig. 2.

Fig.3.

Fig.4.

1. 2. 3. 4. *Hypophyllum virosum.*

Oronge cigue, jaunatre   *Tom. 2, pag. 326, et suiv.*

J. B. Baillière Libraire à Paris.

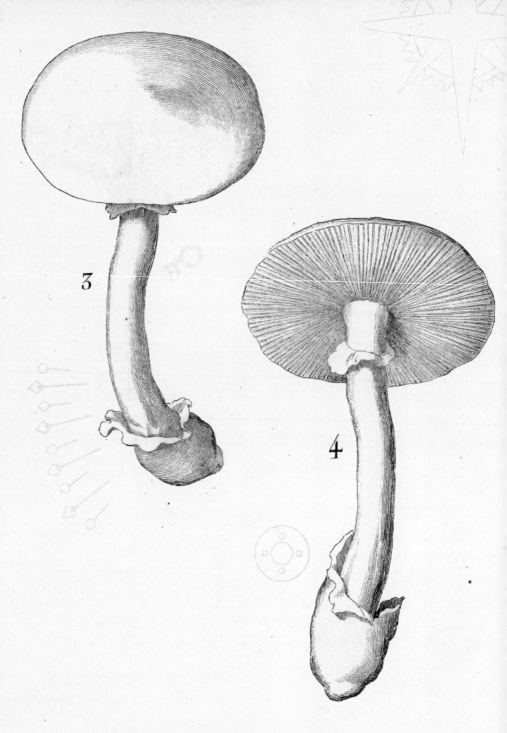

3. 4. Oronge cigue blanche ou du printems, ▲ *Tom. 2.pag. 328.*

Destroying angel

Amanita virosa

Since 1997, inhabitants of the quaint English village of Midsomer have been dropping like flies.

It is down to Detective Chief Inspector Tom Barnaby and his young cousin John to figure out whodunit in Britain's murder capital every week, carried out by a grisly cast of special guest villains wielding ever-more ingenious murder weapons.

If DCI Barnaby could have sneaked a peek at the title of the second episode of *Midsomer Murders'* fourth series, "Destroying Angel" (2001), he would have worked out the cause of at least one of the show's six deaths very quickly indeed. Poor Tristan Goodfellow may have been having an illicit affair with the wife of a murder victim, but no one deserves to have their nice pan of field mushrooms replaced with *Amanita virosa*, the destroying angel.

The bell-shaped *Amanita virosa* is one of a number in the *Phalloideae* subsection of genus *Amanita*, some of the most toxic of all the kingdom fungi. Members of the genus contain varying amounts of phalloidin, phalloin and amanitin, known as phallotoxins and amatoxins, each deeply dangerous. They are particularly unpleasant because, being all-white and, in their youth at least, resembling innocent species, they can be taken for puffballs or button mushrooms. Mature specimens are not unlike open-cup edible species. Nasty symptoms begin between 8 and 24 hours after ingestion and include vomiting, convulsions, watery diarrhoea, delirium, kidney and liver failure, and, in many cases, death.

Antidotes have been trialled, including experiments with *Ganoderma lucidum* (lingzhi) and herbs such as milk thistle, but participant numbers are too small to be considered conclusive. France's Institut Pasteur has developed a serum antidote, but it must be administered soon after ingestion. All in all, it's not worth seeking out one of the only edible members of the family, the blusher or new bride *(A. rubescens)*. It's large, red-brown with cream warts, and "blushes" pink when cut. It tastes good, even if it needs to be thoroughly cooked to be safe, but it just looks too much like the false blusher or panther cap (*A. pantherina*) to make for comfortable consumption by any but the most hardened expert.

As with several really poisonous species of plants and fungi, there is little folklore surrounding the destroying angel. Throughout history, this one has been too important to make up tales about; just teach children to avoid it like the devil. Anyone thought to have ingested *Amanita virosa* should seek immediate medical assistance.

Opposite Destroying angel (*Amanita virosa*) from JL Leveille *Iconographie des Champignons de Paulet*, 1855.

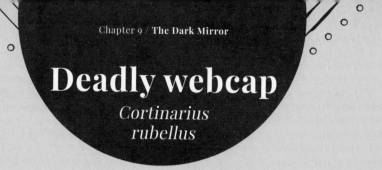

Deadly webcap
Cortinarius rubellus

Such a cute, plump pixie toadstool, with its soft, red-brown cap, wavy edges and solid stipe. So wholesome looking to the untrained eye, just like a chanterelle.

Don't believe this imposter for a moment. Found mainly in coniferous forests, *Cortinarius rubellus* is common in Scandinavia and mainland northern Europe. *Cortinarius* comes from the Latin *cortina* or "curtain", which refers to the partial veil that protects the growing spores in young mushrooms. The veil usually appears as a translucent network of fibres across the gills – hence the term "webcap". *Rubellus* means "red", though this is more the orange-red-brown of a squirrel or fox.

Its cousin, fool's webcap (*Cortinarius orellanus*) is similar but more common in Southern Europe. Both stand out on the forest floor, against a bed of bright green sphagnum moss or dead leaves. Gently undulating, velvety caps, mottled stipes and luscious-looking gills beckon foragers, tempting them to add such dainties to their baskets for a rich supper. A few do.

Generally, the unwary hunter mistakes the deadly webcap for the prized chanterelle (*Cantharellus cibarius*, see page 95). The two mushrooms don't look exactly the same, but the chanterelle is so sought-after, and people are so secretive about foraging sites that newcomers can get overexcited to find a crop "no one else has noticed". In some cases, they do not find out their mistake for two or three days. The flu-like symptoms, usually beginning with headaches and vomiting, may not occur for up to three weeks. If these are not treated quickly, they are fatal.

The cause of the problem – the toxin orellanin – is often difficult to identify due to this delay in symptoms but it directly affects the kidneys and liver. In untreated cases it causes catastrophic kidney failure. Orellanin poisoning accounts for deaths every few years across Europe, and occasionally affects entire villages, such as a case in Bydgoszcz, Poland, in 1952, of 11 deaths.

The most publicized webcap poisoning was that of Nicholas Evans, bestselling author of *The Horse Whisperer*, in 2008. Evans mistakenly gathered webcaps and served "chanterelles" cooked in butter and parsley to his family. Four of them ended up in hospital and three – Evans, his wife and her brother – needed kidney transplants. Evans has been brave enough not to hide his mistake and his experience has re-alerted the public to the dangers of wild mushrooms.

Opposite Deadly webcap (*Cortinarius rubellus*) from Mordecai Cubitt Cooke *Illustrations of British Fungi*, 1881.

8 × 6

M.C.C.

CORTINARIUS (TELAMONIA) RUBELLUS. *Cooke.*
in swamps. Moss, near Carlisle, 1886.

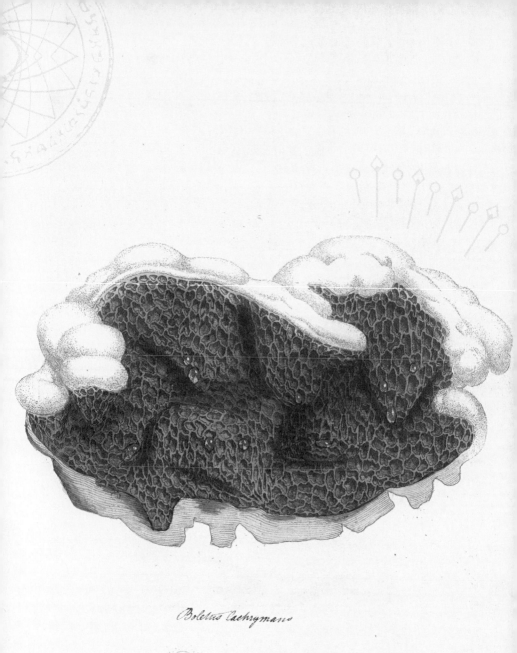

Boletus lachrymans

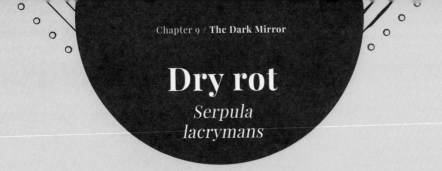

Dry rot

Serpula lacrymans

Serpula lacrymans and its relations have infected buildings, equipment and ships since time immemorial and cost the UK alone around £150 million annually.

Yet despite its name, dry rot does not live in dry areas. It needs both food and water to survive. The famous seventeenth-century diarist Samuel Pepys served as a British naval administrator and, in the 1680s, reported that rotting in ships' timbers was worse than that in buildings. This was some claim, given the amount of wood in the houses of the day. It has been suggested that imported timber, having been stowed in the warm, damp, darkness of ships' holds, not only infected the wood with spores, spreading different strains across the world, but infected the ships themselves.

Serpula lacrymans is native to Northeast Asia, but colonizes anywhere humans take it, via clouds of orange spores. If the spores hit a dry surface, they can lie low for up to 10 years, but under perfect conditions, they go to town, attacking cellulose and breaking it into the sugars they need to grow. The host wood cracks and shrinks into strange, cube-shaped lumps. As the rot sets in, a fluffy white mycelium grows and the fungus produces small, fleshy orange brackets emitting red spores.

The fungus takes the name *Serpula* from the Latin for "serpent", describing the mature fungus's white cords of hyphae, and *lacrymans* ("weeping") from its fruiting bodies bearing water droplets.

Illustrator and naturalist James Sowerby was fascinated by all fungi, and spent some time working out how to treat dry rot. "By my advice, Lord Heathfield caused proper passages for the admission of common dry air which became an effectual cure," he wrote in 1797.

Dry rot can be partially alleviated by ventilation, and the Royal Navy solved their own problem via the invention of iron ships, but *Serpula lacrymans* remains with us. Lawrence Millman notes that it reached epidemic levels in London during the Second World War when the timbers of bombed houses, doused with water, provided a field day for the fungus.

The corners of historic houses are particularly vulnerable, providing the perfect conditions for *S. lacrymans*: darkness, dampness and a ready food supply in old timber. Often taking a hold before it can be detected, the fungus has been called a "cancer". Humans find its breeding grounds – basements, attics, cavities between rooms and under floorboards – difficult to penetrate. Specially trained "rothounds" now sniff out the smallest infestation in a building in return for praise.

Opposite *Serpula lacrymans* (dry rot fungus) from James Sowerby *Coloured Figures of English Fungi or Mushrooms*, 1795–1815.

Two bad caps:
Funeral bell
Galerina marginata

Fool's conecap
Conocybe filaris

Biologically unrelated, both these mushrooms are commonly thought of as hat-shaped: the shallow, head-hugging, skullcap and the pointed, taunting cap of a dunce.

Also known as autumn skullcap, *Galerina marginata's* common name gets straight to the point. Eat the funeral bell and chances are you won't be around to hear your own. The genus name *Galerina* is fluffy enough – it means "like a helmet". Its cap often has a slightly lighter edge, which adds to its helmet-like appearance. Funeral bell is particularly dangerous because it looks so normal; one of those "little brown mushrooms" that seem so similar, they must all be fine to eat.

Common across the northern hemisphere, the funeral bell is a rotter, decaying the dead wood it consumes. Smooth, ranging in colour from pale yellow or brown to dark orange, it has a creamy underside and crowded gills, though it becomes slimy when wet and can look sickly. There's a reason for this. *G. marginata* contains the same amatoxins as death cap (*Amanita phalloides*) and destroying angel (*Amanita virosa*). Humans – and animals – who have accidentally ingested the fungus may experience vomiting, diarrhoea, hypothermia and catastrophic liver failure.

Conocybe filaris is one of those slender little lawn mushrooms, with a cute, conical hat in homely brown, and a skinny stipe that looks hardly capable of supporting its own weight. Halfway down the stem, a little frilly skirt makes it even less dangerous-looking. The shape – a bit like a dunce's cap when young – should lend a clue to its common name: fool's conecap. Yet another rotter, it can be found on woodchips or compost heaps, busily decomposing matter, a very useful trait. Like the deadly skullcap, its colour can change depending on how wet or dry it is, but the cap is usually brown and so thin it is almost translucent. It often has an umbo (raised centre) and, as it gets older, the mushroom becomes flat, stretching out to spread its spores.

Both these mushrooms are dangerous for their sheer ordinariness. They are common and bland. As with so many of the Earth's most deadly killers, few mess around romanticizing them with fairy tales and campfire novelties. Debutant foragers would be wise to acquaint themselves with the bell and the fool.

Opposite Figure B: fool's conecap (*Conocybe filaris*) from Mordecai Cubitt Cooke *Illustrations of British Fungi*, 1881.

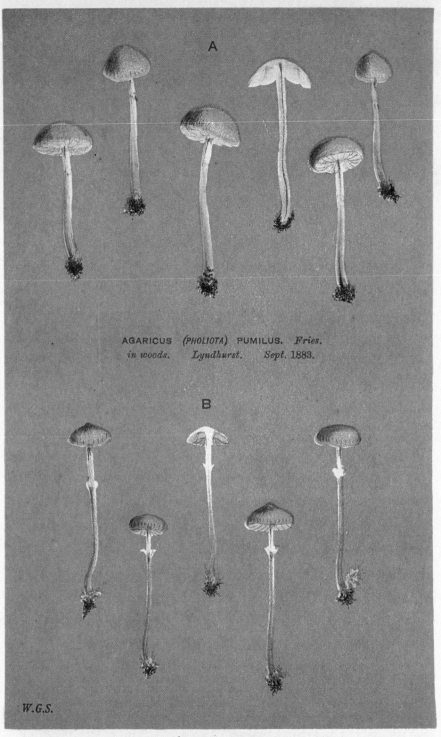

A

AGARICUS *(PHOLIOTA)* PUMILUS. *Fries.*
in woods.   Lyndhurst.   Sept. 1883.

B

W.G.S.

AGARICUS. *(PHOLIOTA)* MYCENOIDES. *Fries.*
amongst moss, in swamps, &c.

Poison fire coral

Podostroma cornu-damae

Its deep red stalks branching and dividing, this fungus looks more like it should live with exotic corals on the seabed than on the forest floor.

Some believe it would be safer there, too, though the toxicity of the poison fire coral is currently a hotbed of mycological debate.

In Japan and Korea, the *kaentake* ("fire mushroom") has been treated with extreme caution for centuries, considered a seriously dangerous fungus that should not even be touched. Its reputation as a killer grew as it was discovered in countries such as China, Papua New Guinea and Thailand, but when it was also recorded in Australia, the internet broke. The resulting rash of articles ranged from seriously scientific to clickbait, the latter starting a whole new folklore of the web.

It is true that *Podostroma cornu-damae* is not good for humans. While its bright, salmon-pink branches are pretty, its cocktail of trichothecene mycotoxins are even said to be capable of being absorbed through the skin, causing irritation and swelling. Eating it is really not a good idea. Medical journals describe case studies including stomach pains and vomiting, peeling skin, hair loss, multiple organ failure, necrosis (the death of body tissue), respiratory problems that required artificial ventilation and eventual death. The press went wild at another symptom: the shrinking of the cerebellum. Literally, the fungus was said to shrink your brain. This led to speech impairment, hallucination and difficulty moving.

Recently, the tables have begun to turn on *P. cornu-damae*. Even as the press went crazy over the Australian discovery of the "killer fungus", a quiet mycological backlash began batting for the enemy. The anonymous Japanese author of konokobito.com argues that the poison fire coral cannot be a "killer mushroom" because it does not lie in wait to attack humans, it just happens to react biologically with the humans that interact with it. The author says that photographs purporting to be the horrific results of touching the mushroom actually depict someone suffering from athlete's foot. While admitting people should not touch their eyes after touching *P. cornu-damae*, the author suggests it is nowhere near as deadly to touch. They think the hysterical tone of some articles detract from the message of respect over revulsion.

Poison fire coral is *not* safe to eat, but it performs a vital part in ecology. For all its dangers to the humans that choose to touch it, *Podostroma cornu-damae* is vital to life. The same must be said for all such fungi .

Opposite Poison fire coral (*Podostroma cornu-damae*) illustration by Malcolm English, 2022.

Deadly dapperling

Lepiota brunneoincarnata

The Latin name of this toxic fungus comes from lepis, *meaning "scaly" – the same root as leprosy – and* ot, *or "ear".*

*L*epiotas tend to have scales radiating across their caps, either in concentric circles or overlapping, like little brown shingles on a white roof. The deadly dapperling has crowded, creamy white gills and a pink-brown stipe covered in fibrous scales. It grows in fields, parks and gardens, where edible mushrooms are also found. *Lepiota cristata*, the stinking dapperling, is also poisonous. While it looks a lot like the edible *Macrolepiota procera*, and any number of the Agaricaceae mushrooms, it also smells terrible, providing a clue to hunters that it is none of these. The scales of *Chlorophyllum rhacodes*, or shaggy parasol, are more pronounced and sometimes resemble short, matted white hair. Some people eat it, but it may cause nausea and is best avoided. *Lepiota felina*, the cat dapperling, has markedly concentric rings of scales leading into a dark brown central "eye" – this one's not edible either.

Like many poisonous fungi, despite – or perhaps because of – its commonness across the world, there is little folklore to romanticize the dapperling. The modern games' world is one of our most fruitful sources for new legends – here, the deadly dapperling may at last have its moment in the sun. The hugely popular *World of Warcraft* series often builds its fantasy worlds on traditional real-world sources. Glutharn's Decay is an area within Maldraxxus, the citadel of the Necrolords, where may be found the slime creatures used by the House of Plagues to create potions. It's also the haunt of the Deadly Dapperling, a Fungarian (a race of mushroom people) who has a bone to pick with the House of Plagues, which habitually imprisons and kills Fungarians. Deadly Dapperling is loosely based on *L. brunneoincarnata*, much as Princess Penicilla and Humon'gozz are based on *Penicillium* and *Armillaria gallica*.

In many ways, this is the way folklore has always evolved. For millennia, people have taken the basics of their own world and created stories around them. In this case, *World of Warcraft* has taken aspects of the fungal kingdom – habitats, growth patterns, properties and uses of real mushrooms – and new lore has been created within an imaginary realm. The setting may be a fantasy roleplaying game rather than the campfires and hearths of old, but the system is the same. Fungarians are not so different from, perhaps, Japanese *yōkai* (spirits) or the humanoid toadstools from Russian fairy tales, they just live in a digital Shadowlands rather than a magical forest. Humans will never cease telling themselves stories, however virtual their world.

Opposite Deadly dapperling (*Lepiota brunneoincarnata*) by Malcolm English, 2022.

The Paris Poisoner

Many "mushroom murders" through history have been difficult to prove, not least because the poisoner's weapon of choice, Amanita phalloides, *has a delayed reaction that can deflect attention from the victim's "last supper". In one case, however, the evidence is clear.*

If it hadn't been for the fact that Europe was at war when he was serially dispatching his "friends", Henri Girard's name might be more notorious. Then again, a second fact – that many of his potential victims survived to testify – may just mean he wasn't as good at murder as he thought.

Admittedly, mushrooms were not Girard's favourite weapon. Born in 1875 and dishonourably discharged from the Hussars in 1897, the petty thief, gambler and swindler originally dabbled with typhus bacteria. Around the same time, he also noticed how lax the regulations were around personal insurance.

In 1910, he ingratiated himself with an insurance agent, Louis Pernotte, who allowed his new friend to take out a policy whereby, should one of them die, the other would be the beneficiary. Girard then took out another policy with another company. Three more followed. In 1912, the Pernotte family became ill after dining with the Girards. All recovered except Louis, who had allowed Girard to later inject him with "camphorated camomile". He died of cardiac arrest. Girard collected the insurance.

In 1913, Girard took out six insurance policies for an old friend, M. Godel. Godel went to dinner with the Girards, then came down with typhus. He recovered. 1914 saw Girard secretly insuring the life of an "M.

Delmas", with himself as beneficiary. Again, his victim survived but became suspicious and removed himself from Girard's company. The next life secretly insured was that of a Mimiche Duroux; he was also invited to dinner and also didn't die.

Girard needed to up his game. When he met a Mme Monin in 1918, he and his lover Jeanne Droubin took out three different life insurance policies on the young war widow. Girard invited her for cocktails; she lost consciousness on her way home and later died. Two insurance companies paid out, but the third didn't buy a healthy young woman suddenly dropping dead and started investigating.

Mme Monin's autopsy revealed Girard's modus operandi was now *Amanita phalloides*. Girard had been brazen enough to mention mushrooms, dinner invitations and his victims' symptoms in his diaries.

In the 1950s, Gordon Wasson became fascinated with the case, not because Girard used the death cap as poison but because it appears that, had he not trusted an overcautious fungi identification manual, the body count would have been much higher. Girard obtained his mushrooms from le Pere Theo, an elderly hobo who collected them in the forest of Rambouillet. The manual warned that nearly all *Amanitas* were deadly, so Girard accepted several varieties that

are not poisonous, inadvertently creating one of the only known examples of mushroom misidentification leading to lives being saved.

Alas, we will never know if this theory is true. In 1921, after three years of evidence gathering and a couple of exhumed bodies, Girard was sent on remand to Fresnes Prison in Paris. There, he cheated the French courts of finding him guilty by ingesting a germ culture he had smuggled in. His wife and mistress both received life sentences.

Below Death cap (*Amanita phalloides*) by Elsie M. Wakefield, Kew Collection, 1915–45.

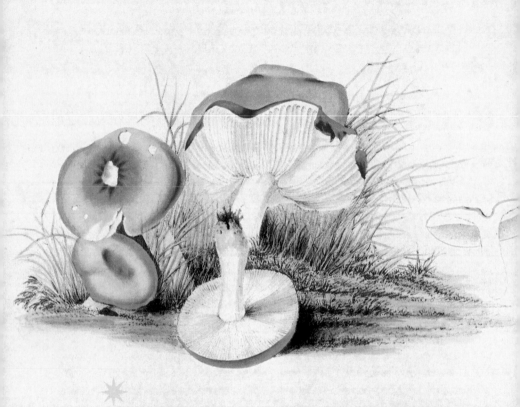

Agaricus emeticus. *Schœffer*.

The pretty little sickener

Russula emetica

Anything with the word "emetic" in its Latin name isn't likely to be recommended as a digestif.

Yet *Russulas* are much appreciated in Northern and Eastern Europe and, for those who know which of the family have heat-sensitive toxins that can be killed by thorough cooking, they are a treat. The whole group should be treated with caution as everything about them, from looks to smell, is topsy-turvy.

For example, the frankly queasy-looking quilted green russula (*R. aeruginea*) is said to have a fine flavour – if cooked properly. Hintapink, or bog russula – *R. paludosa*, blush-pink with a darker, slightly sunken centre – smells terrible but is perfectly edible. The wine-dark *R. xerampelina* (shrimp russula) starts out with a slight odour of boiled shellfish and matures to a deep fishy pong, but is considered a fine table mushroom when fried or used in soups. The sickener (*R. emetica*), however, will lure an unwary forager with its deep, rose-red cap, snow white gills and rich fruity scent, only to make them throw up after they've tasted it. Unless consumed by a very small child or someone with underlying health issues, it's unlikely to kill but, eaten raw, the mushroom can induce nausea, vomiting, stomach cramps and diarrhoea. Despite this, in parts of Eastern Europe, the colourful skins are carefully peeled away and used for flavouring goulash (the rest is, wisely, discarded). Usually found in pine forests, in a rich bed of bright green sphagnum moss, the optional "pretty" part of the sickener's common name is well deserved, an opinion shared by the red squirrels who gorge on it without ill effect.

The Kuma tribe of Papua, New Guinea, tell how the first man on Earth fashioned a mushroom, which he threw into the sky to make the Moon. The first woman threw her mushroom to form the Sun. One of the tribe's rituals is a "shivering madness" brought on by boiling and consuming various *Russulas* and *Boletus*. Everyone partakes, but only a few are affected by *komugl tai*, says Robert Rogers, a state that roughly translates as "shivering bird of paradise, deaf to reason". This may be due to some kind of hereditary susceptibility. *Russulas* are widely used in folk medicine. TCM's famous tendon-easing powder uses several dried mushrooms, including *Russulas*. Some of the short-stalked varieties are thought to contain antioxidants and antimicrobial properties and, increasingly, medical science is investigating possible uses for conditions as diverse as HIV-AIDS, malaria and certain cancers.

Opposite *Russula emetica*, the sickener, from Anna Maria Hussey *Illustrations of British Mycology*, 1847–55.

Dead man's fingers

Xylaria polymorpha

Uncannily like the rotting flesh of warty, swollen fingers straining from the bare earth, the fruiting bodies of Xylaria polymorpha *seem to be making one last-ditch attempt to escape from a dead man's grave.*

It doesn't help that *Xylaria polymorpha* usually appears in clumps of digit-like structures, short and bloated, bent like ancient, arthritic knuckles. Sometimes they're pale with blue-black staining, sometimes long and skeletal; mature "fingers" look like they're wearing heavy-duty, black rubber gloves. Internet photographers vie to collate the creepiest-looking examples, whether blue-grey with whitish tips, or black with disturbing, flesh-pink tops.

This genuinely spooky-looking fungus performs a useful function. It grows on the buried dead wood of broadwood trees, especially beech, breaking down the matter into a form that it can ingest. Because it feeds on polysaccharides (the compounds that bind the main components, cellulose and lignin, into solid wood), *X. polymorpha* leaves behind a soft, easy-to-eat and nutritious meal for forest feeders. It is less welcome in a garden – *X. polymorpha* may settle on damaged roots or bark, causing the tree to rot. Apple can be susceptible. Infected trees should be removed quickly, as they are liable to sudden collapse.

Perhaps surprisingly, given its Halloweenish feel, dead man's fingers is relatively short on folklore. In Lithuania, these mushrooms were said to be the fingers of Velnias, the Baltic god of the dead. Velnias comes third in the pantheon of Baltic gods, after Dievas, the sky god, and Perkūnas, god of thunder. Although he raises whirlwinds and leads hosts of the dead, Velnias was originally considered a good god, guarding the treasures of the earth, protecting the poor and punishing the wicked. His fingers were said to reach out from his dead world to feed the poor. Alas, he fared badly in Christian times, when the god of the underworld became equated with the devil.

Dead man's finger is not generally considered to be edible, though it is unlikely many would be tempted to try it. It does have some folk uses, however. In Indian Ayurvedic medicine, women take them with milk to promote lactation. A relation, *Xylaria nigripes*, which has whitish fruiting bodies, is used in TCM for a number of conditions and is currently being investigated for possible antidepressant effects in patients with epilepsy.

Another relation, *Xylaria longipes*, dead moll's fingers, is smaller, duller, wartier and with a more pronounced stipe. It is mainly found on the stumps of dead beech and sycamore. *X. longipes* is not edible either, but children are even less likely to want to put this creepy customer in their mouths.

Opposite Dead man's fingers (*Xylaria polymorpha*) from James Sowerby *Coloured Figures of English Fungi or Mushrooms*, 1795–1815.

Clavaria digitata

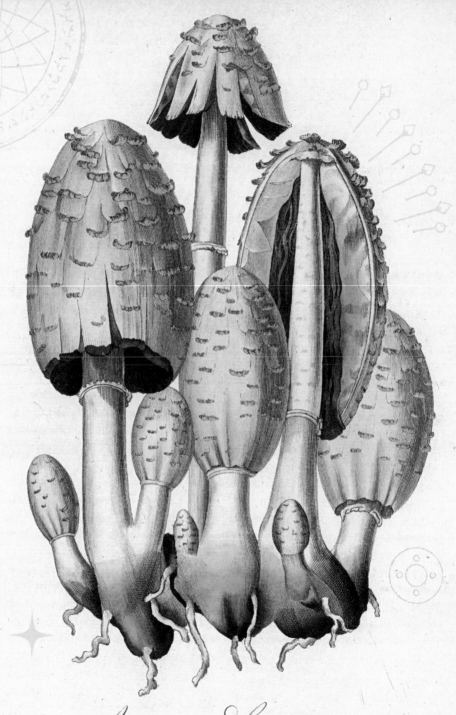

Agaricus fimetarius

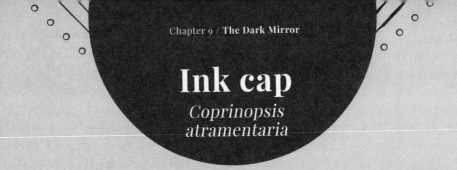

Ink cap
Coprinopsis atramentaria

A large family of delicate, fragile mushrooms, the ink caps are both beautiful and short-lived, often lasting just a few hours. "Deliquesce" is a word meaning "to dissolve".

When used in a fungal context, it describes the actions of a mushroom that liquefies its gills to distribute spores. The common ink cap (*Coprinopsis atramentaria*) does release spores aerially, into the atmosphere, but also commits fungus suicide to distribute the rest. Across a few hours, these beautiful, pale brown, black-edged bonnets on spindly white legs become victims of their own enzymes. Their whitish gills turn grey, then black, before dissolving entirely into a puddle of brown-black fluid. Turnover is fast; the mushrooms huddle together in colonies, a generation often living and dying in a day.

Eaten alone, it's unlikely *Coprinopsis atramentaria* will do much (short-term) damage. Alternative folk names, however, such as alcohol inky and tippler's bane, hint at a darker potential problem. The compound coprine is found in most ink caps and can usually be broken down by the body with a combination of carbon dioxide and water. If you drink alcohol shortly before (and up to five days after) eating *C. atramentaria*, the mushroom inhibits the action of the liver. Toxins then accumulate in the liver, making the patient very ill. Some ink caps, including *C. nivea*, *C. narcotica* and *C. radicans*, have psychoactive properties.

"Coprinus" may be from the Greek word *kópros*, meaning "excrement", but it is not particularly apt, as many ink caps in this and related genera prefer decaying wood to dung. Their fruiting bodies can be found by tree stumps or on roadsides and, for such dainty beings, wield considerable strength, pushing through layers of frozen ground. The shaggy ink cap (*Coprinus comatus*) does prefer dung or rich soil, and its strength may be enough to push through layers of tarmac in roads, causing potholes. *C. comatus* suits another of its common names, judge's or lawyer's wig, thanks to the curling white scales over the cap. It is edible and best harvested young, when it looks like a white, scaly lollipop.

Ink caps are used in TCM as an anti-inflammatory; historically, they were used as treatment for burns in Sweden. Once again, science is researching them for possible treatment of some cancers. Their liquid used to be boiled up with cloves and water or urine to make a sepia-toned ink, fixed with iron filings. Recently, artists have been experimenting with fungus ink again, sometimes placing an entire fruiting body onto paper or cloth to create spooky, mushroom prints.

Opposite Common ink cap (*Coprinopsis atramentaria*) from William Curtis *Flora Londinensis*, 1775–98.

Zombie ant fungus

Ophiocordyceps unilateralis

One of the most fundamental human fears is the loss of free will; to have one's mind and body controlled by another, up to and after death. In the insect world, this is only too possible.

The *Ophiocordyceps* genus comprises a group of entomopathogenic fungi – organisms that parasitize insects – feeding from, living in, and usually killing their host. Caterpillar fungus (*Ophiocordyceps sinensis*) is a well-known member of the genus (see page 146) but there are several similar fungi, including *O. unilateralis*, mainly found in tropical regions. All manner of different ants fall prey to this "zombie-maker" though *Camponotus leonardi* (carpenter ants) are the most common. Carpenter ants usually live high in the canopy but need to come down to move between trees, exposing them to spores.

Once infected, the ant's "zombie" behaviour always follows the same pattern. It climbs up a plant to a height of about 26cm and clamps its mandibles onto the vein of a leaf or the underside of a twig. By now, the ant has no control over its body, which usually slides from the leaf, hanging by its jaws in a death grip. Now secured, the fungus has no further use for the living insect. It gradually kills its host, then lives on it, invading the soft tissue and strengthening the exoskeleton, which it still needs – for now. It sends mycelia out from the body to secure it further, while emitting antifungal secretions to kill off other fungi seeking a share of the spoils. After 4 to 10 days, it is ready to spore, and sends a long, thin fruiting body

out of the dead ant's head, covered with "perithecia". These ovoid, spore-bearing chambers, just beneath the surface of the fruitbody, release a cloud of spores through pore-like openings that rain onto the forest floor – and the next victims. The discarded corpse falls to the ground to join its unlucky compatriots.

For years, *Ophiocordyceps unilateralis* was thought to eat the ant's brain. However, researchers at Penn State University discovered that the fungus controls the ant in a more holistic way, attacking the head, thorax, abdomen and legs at the same time, forming a 3-D network to control its movements. Indeed, it keeps the brain alive long enough to commit that last death grip. This makes the fungus more necro-puppeteer than zombie mind-controller.

The explanation does not seem to have comforted game designer Neil Druckmann. After seeing the BBC documentary *Planet Earth*, which explored *Ophiocordyceps*, he created *The Last of Us*, a zombie-apocalypse video game where a mutant insect-eating fungus jumps species to humans, controlling their minds beyond death. It's now an HBO TV series. Whoever said folklore was a thing of the past?

Opposite *Ophiocordyceps unilateralis* sporing body growing out of infected dead ant, by Katie Scott, from *Fungarium*, 2019.

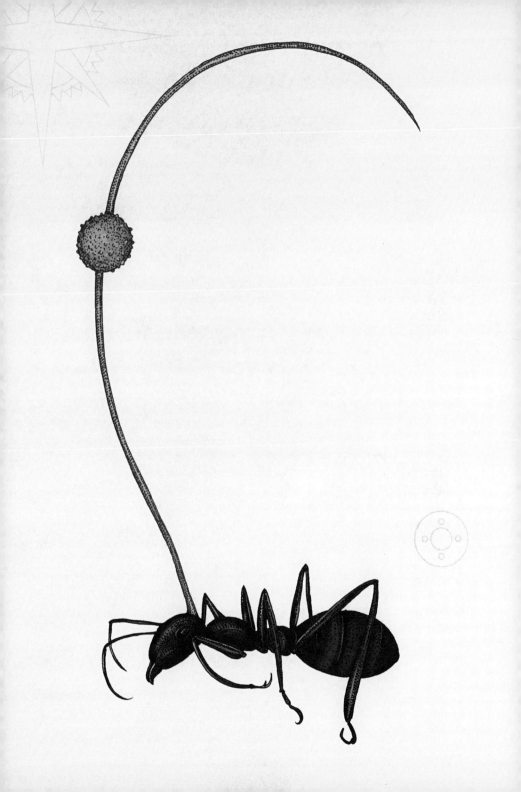

PL. LV.

1

2

3

Fig 1.... *Hypophyllum xerampelinum*.

La Feuille morte. ◯ *Tom.2.Pag.148.*

Fig 2.3. *Hypophyllum quinque partitum*.

Le Champignon cinq parts. ◯ *Tom.2.Pag.148.*

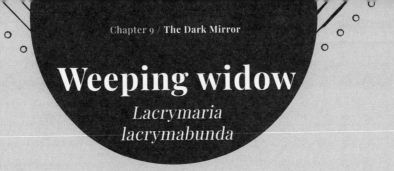

Weeping widow

Lacrymaria lacrymabunda

Haunter of grasslands, woodlands and park verges, this "widow" is often found "weeping" beside recently dead trees.

A medium-sized mushroom in a rounded cone shape, it starts life as a jolly spherical blob. As the fruiting body matures, the stipe becomes pronounced, and the umbo (raised centre) slowly flattens. The "widow's" cap has a fibrous texture, as though "she" is wearing a thick woollen shawl. In damp or wet weather, the mushroom "weeps" inky-black droplets, stained by its spores.

Lacrymaria means "tears" in Latin; this was what stuck when naming this teardrop mushroom. It calls to mind *pleurants* ("the weepers"), mourning statues, usually female, that "pray" or "cry" over tombs. A medieval tradition, weepers enjoyed a renaissance in nineteenth-century graveyards such as the Parisian Cimetière du Père Lachaise and the "magnificent seven" cemeteries of London. As they became overgrown, the figures took on a new folk resonance as unintended omens of darkness.

In America, individual memorials attracted their own legends. The art-deco woman weeping over the 1929 tomb of Laura Daniels Fereday in Spanish Fork, Utah, is said to cry real tears if approached by someone with their eyes closed. Another weeper haunts the statue of escapologist Harry Houdini in Machpelah Cemetery, Queens, New York. *Pleurants* have menaced horror movies since the 1950s,

appearing most recently as *Doctor Who* villains.

While most modern stories of weeping widows wallow in tragedy, it wasn't always so. Traditional tales, from Europe to China, dating as far back as Aesop, poke ribald fun at the ability of some to "get over" their grief when a handsome new suitor arrives.

Less benevolent is the South American La Llorona, a female spirit denied entry to heaven who floats about in her funeral shroud, weeping inconsolably. Depending on the storyteller, her crime was either drowning herself or her children after witnessing her unfaithful husband. Now she is out for revenge. Various versions, including five movies, see La Llorona kidnapping children or avenging infidelity in suitably bloody fashion.

The weeping widow mushroom is not so deadly. Some even consider it edible, but it has a bitter taste and needs cooking, upon which it quickly turns mushy. As a great favourite of flies, it's worth checking over carefully for larvae. Like many "wayside" mushrooms, weeping widows should not be picked near busy roads, as they easily absorb pollution. Come to think of it, maybe this tragic mushroom should be left to her tears after all.

Opposite Figure 1, weeping widow (*Lacrymaria lacrymabunda*) from JL Leveille *Iconographie des Champignons de Paulet*, 1855.

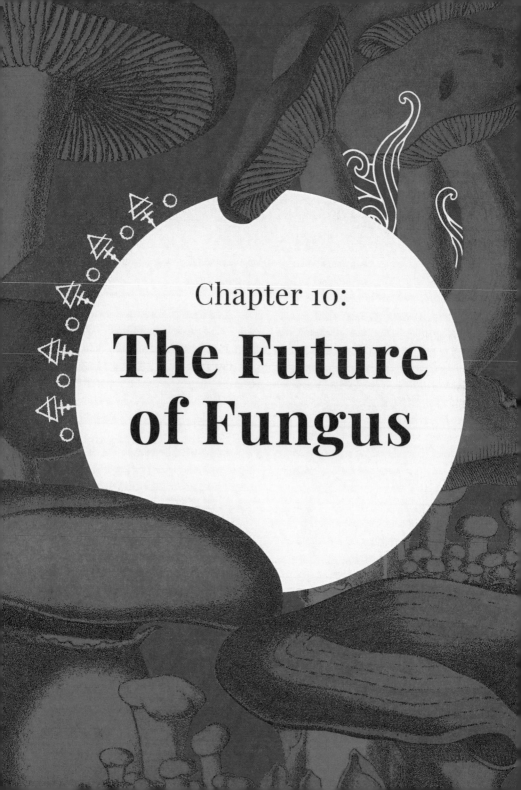

Chapter 10:
The Future
of Fungus

In 1991, scientists were astonished to find fungus growing inside the melted reactor at the site of the 1986 Chernobyl nuclear accident. Far from just surviving, *Cladosporium sphaerospermum* was thriving. The fungus was, it turned out, radiotropic – able to convert the leaked radiation into energy – and NASA scientists are now investigating it as a possible radiation shield for astronauts on a projected mission to Mars. It is just one of a range of mind-boggling potential uses now emerging for the least-understood kingdom on Earth.

In the last few decades, new technologies have been developed to study the fungal world, inspiring scientists – and others – to think creatively about the kingdom and the problems humanity faces.

It could be said this is only a twenty-first-century reinterpretation of what our ancestors did, observing fungi then working out what they could be useful for. It helps that we have discovered one fascinating fact: fungi are more closely related to animals than plants. In 1987, English zoologist Thomas Cavalier-Smith came up with a new word, "opisthokonta", to describe a group of animals and fungi that share properties. For example, both humans and fungi breathe oxygen and exhale carbon dioxide. Both can also suffer from biological pathogens. Some fungi have evolved powerful antibacterial, antibiotic and antiviral properties that we may be able to harness. Indeed, humans already did. Every time ancient Egyptians used mouldy bread to dress a lesion or Native Americans packed out a wound with the spores from a puffball, they were exploiting fungus's self-defence mechanisms. Similarly, we are only scratching the surface of the many anticancer agents that may be found in various fungi. Other conditions that may find lasting benefits from fungal research include diabetes, nerve problems, cardiovascular disease and immunosuppressive issues. While investigations are in their early stages, the results so far are impressive. Research in other areas includes long-term strategies against diseases in plants, including biocontrol of insects and nematodes, weeds and even fungal conditions such as rusts and smuts.

In 2014, a demonstration tower called Hy-Fi, by design studio The Living, was displayed outside the Museum of Modern Art in New York. The structure used fungus-based bricks designed to compost themselves after their useful life. Once baked, the fungus stops growing, keeping the building's shape, though it is possible it could be merely paused in its growth. It could then be rehydrated to heal any future damage or wear and tear, or merely composted when no longer required. Recent experiments show that mycelium can conduct electrical signals, introducing the intriguing possibility of a house being able to react to its environment, a little like a brain. Inside, fungus-based textiles can "grow" anything from vases, lampshades and chairs to bedroom slippers.

Previously, the closest fungus got to fashion was the 1980s craze for "puffball" skirts – an unflattering garment gathered at the top and bottom to create a short, baggy balloon of fabric around the waist. Now, the fabric itself can be fungal. Material from mycelium is waterproof, hypoallergenic, non-toxic and fire-resistant. It can be thin, almost papery, or tough and durable, as with the Mylo Unleather fabric made by American company Bolt Threads and used extensively by designer Stella McCartney.

In some ways, using fungi as food almost seems too obvious – from cave people consuming mushrooms to yeast products such as Marmite, invented in 1902 – it has been going on for years. The story of Quorn begins in the 1960s, when food shortages led British company Rank to look into the *Fusarium venenatum* fungus as a possible alternative

Opposite Reddish-brown crust fungus (*Hydnoporia tabacina*) from James Sowerby *Coloured Figures of English Fungi or Mushrooms*, 1795–1815.

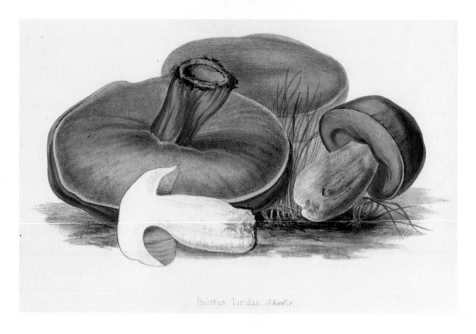

Boletus luridus. Schaeffer.

protein source. Mycelium's mass of thread-like fibres can be manipulated so that laboratory-created fake meat can look, taste and smell like the real thing. Even the texture can resemble muscle. Convincing fake chicken breasts and even steaks are now possible, using different amounts of added fat to mimic each cut. The product is cultivated in tanks, and takes a fraction of the time, space and raw materials needed to rear an animal. Unlike plant-based alternatives, fungal meats can be flavoured from scratch rather than having to first conceal the original taste of, for example, soy or legumes.

Even going back to basics can reveal new information about this still largely unexplored kingdom. When mycologists at the Royal Botanic Gardens, Kew DNA-tested a supermarket pack of dried porcini mushrooms, they discovered three different species of fungi, each new to science.

Work is also being done on fungal biofuels.

A team led by Gary Strobels at Montana State University has been studying a fungus growing in the ulmo tree (*Eucryphia cordifolia*) of northern Patagonia. Research underway shows that *Ascoryne sarcoides* produces hydrocarbon molecules similar to fossil fuels. Critics point out that, currently, "mycodiesel" has downsides in carbon emissions, possible food shortages and potential deforestation issues, but further research may yet find greener alternatives. There is almost certainly a whole host of uses that even fungal left-fielder Paul Stamets hasn't dreamed of yet. Hopefully such discoveries will squeak through just in time to save the planet.

Above Lurid bolete (*Suillellus luridus*) from Anna Maria Hussey *Illustrations of British Mycology*, 1847–55.

Opposite Straw mushroom (*Volvariella volvacea*) from James Sowerby *Coloured Figures of English Fungi or Mushrooms*, 1795–1815.

Since the beginning of humanity, there has been one aspect of fungi that overrides fascination, curiosity and passion: poison.

Even the most rampant mycophile has profound respect for the kingdom – or at least, the mycophile that is still alive does.

Knowing the difference between nigh-on identical mushrooms can be the difference between a delicious meal and death. In some cases it can mean both. The death cap (*Amanita phalloides*) can be very nice while it's being eaten. Antidotes are just as hit and miss. Talking to the *Sydney Morning Herald* as recently as 2020, mycologist Dr Teresa Lebel was only half joking when she suggested anyone foraging for mushrooms should "leave half for the coroner". Bringing a sample of the wild mushroom that someone taken poorly has just eaten can help a doctor work out how to treat them.

There is no shortcut to learning which mushrooms to consume and which to avoid, but that does not mean that folklore hasn't tried to make the process easier. Tall tales of how to tell a good mushroom from bad range from the ridiculous to the "possibly something behind it but really ... don't".

None of the following superstitions mean a mushroom is edible:

- If it does not taste bitter
- If it is gathered under a full moon
- If insects do not avoid it
- If it is white
- If it is plain coloured
- If it grows on wood
- If it has tooth marks in the cap
- If it grows in an orchard where apple trees are blooming
- If the gills are not pink

- If it is not brightly coloured
- If its stalk splits easily
- If it is not porous

Poisonous mushrooms will *not* turn blue or black (or vice versa) if touched by the following:

- A silver spoon
- A silver coin
- Onions
- Salt

Neither will they necessarily turn parsley yellow or separate milk. Cooking with aubergines will not remove any poison present, and peeling the cap is not a reliable method either.

There really is only one way to avoid mushroom poisoning: learn your fungi. Mycophile chefs make a sensible recommendation: unless you personally know the chef's fungal credentials, be wary of the special on a restaurant menu that offers freshly gathered wild mushrooms. The same goes for interesting-looking purveyors of hand-gathered wild mushrooms at markets.

The Tonga people of Zambia are big fungi fans, often preferring nothing more elaborate than boiling them with onions, tomatoes and groundnuts. They dry their mushrooms, fry them and roast them, the women and girls gathering them with expert eyes in the early mornings. They have a saying, however: *sibbuzya takolwi bowa*, meaning "The one who asks is the one who does not get poisoned by mushrooms". In other words, if in any doubt, ask. Folklore is at its best when it doesn't kill you.

16-9-86

agaricus laccatus.

Bibliography

Trying to list all the books, articles and papers consulted for *The Magic of Mushrooms* would be impossible in a short space like this, but here is a selection of the works that I found most consistently useful:

Merlin Sheldrake's *Entangled Life: How Fungi Make Our Worlds, Change Our Minds and Shape Our Futures* (The Bodley Head, 2020) became an instant classic pretty much the second it hit the bookstores in 2020. The ride is both thrilling and astounding, told with honour and poetry.

The indomitable **Wassons, Robert Gordon and Valentina**, stand proud for their nigh-on definitive work, *Mushrooms, Russia and History, Volumes 1 & 2*. Dedicated, detailed and passionate, the book was written in 1957 and an original copy will set you back several thousand pounds. I am very glad to say it is easily available online as a PDF. **Robert Gordon Wasson's** *The Death of Claudius or Mushrooms for Murderers* (Botanical Museum Leaflets, Harvard University, 1972) was also extremely useful.

Without **Robert Rogers's** encyclopaedic *The Fungal Pharmacy* (North Atlantic, 2011) this book would be infinitely the poorer. The depth and breadth of Rogers's research is truly astounding, his writing style eminently readable, his love for all fungi captivating.

The sheer vivacity of *Fascinated by Fungi: Exploring the Majesty and Mystery, Facts and Fantasy of the*

Quirkiest Kingdom on Earth (First Nature, 2011) is a joy to read. At just shy of 500 pages, it has clearly been a labour of love for **Pat O'Reilly**, beautifully illustrated and hugely detailed.

The Royal Botanic Gardens, Kew's own *Fungarium* (Big Picture Press, 2019), curated by **Katie Scott** and **Ester Gaya**, is not only clear and precise but an object of beauty in its own right.

Patrick Harding's *Mushroom Miscellany* (Collins, 2008) is a hugely enjoyable and informative read. **George W. Hudler's** *Magical Mushrooms, Mischievous Molds* (Princeton University Press, 1998), is a fine primer in fungal processes. *Lichens, Naturally Scottish* by **Oliver Gilbert** (Scottish Natural Heritage, 2004) gives a passionate account of why we should all take more notice of lichen and what we are doing to it. The delightful *Fungipedia*, by **Lawrence Millman** (Princeton University Press, 2019) is filled with fascinating fungi facts, including a welcome dose of mythbusting.

The Sacred Mushroom and the Cross: A Study of the Nature and Origins of Christianity Within the Fertility Cults of the Ancient Near East by **John M. Allegro**, (Gnostic Media Research, 2009) delves deeply into dark corners of fungal history, with interesting – if decidedly oddball – conclusions.

I consulted many, many books on general folklore, from regional stories and calendar customs to worldwide superstitions. Fungal folklore often comes in dribs and drabs, from old books, gazetteers and even maps. **Frank Dugan's** paper *Fungi, Folkways and Fairy Tales: Mushrooms & Mildews in Stories, Remedies & Rituals, From Oberon to the Internet* (*North American Fungi*, January 2008) is a superb piece of modern-day folk research that gathers together much of what is out there; the rest was picked up bit by bit, tale by tale, superstition by superstition, from a range of sources.

I did not focus on any of the exhaustive foraging companions, as this book is more about the

stories behind the fungi than attempting to be an identification guide, but **Rich Wright**, a mycologist at Kew, to whom I am in great debt, recommends Paul Sterry's *Collins Complete British Mushrooms and Toadstools* (2009) as a good general guide, or a guide relevant for your locality.

Previous page *Inocybe geophylla* var. *lilacina*, labelled *Agaricus laccatus*, drawn by Florence H. Woodward, Kew Collection, late-nineteenth century.

Below Parrot waxcap (*Gliophorus psittacinus*) from Anna Maria Hussey *Illustrations of British Mycology*, 1847–55.

Index

Picture credits

The publishers would like to thank the following sources for their kind permission to reproduce the pictures in this book.

All other images are taken from the Library and Archives collection of the Royal Botanic Gardens, Kew.

ALAMY STOCK PHOTO: Album: 23, 74, 85; Artokoloro: 119; Chronicle: 136; Keith Corrigan: 124; Everett Collection, Inc.: 18; Heritage Image Partnership Ltd: 15, 27; Interfoto: 12, 139, 160; Matteo Omied: 89; Jonathan O'Rourke: 125; Pictorial Press: 129; Science History Images: 54; sjbooks: 109; UtCon Collection:81; Ivan Vdovin: 121; Walker Art Library: 90-91

ARMITT LIBRARY & MUSEUM CENTRE: 105

BONNIER BOOKS UK: 49, 147, 189 Katie Scott

BRIDGEMAN IMAGES: Christie's Images: chapter openers

GETTY IMAGES: bauhaus1000: chapter openers; Dea Picture Library/De Agostini: 56-57

PUBLIC DOMAIN: 22; /Arthur Rackham: 104

SHUTTERSTOCK: Cci: 68; Grainger: 106; Kseniya Parkhimchyk: chapter openers; Bodor Tivadar: 165

Every effort has been made to acknowledge correctly and contact the source and/or copyright holder of each picture. Any unintentional errors or omissions will be corrected in future editions of this book.

The Kew Publishing team would like to thank the following for their help with this book; Rich Wright, mycologist and outreach officer, Plant and Fungal Trees of Life Project; Lee Davies, Kew Fungarium collections curator; Robert Bye, Universidad Nacional Autónoma de México; Craig Brough, Julia Buckley, Patricia Long, Anne Marshall, Cecily Nowell-Smith and Lynn Parker from Kew's Library and Archives; Paul Little for digitisation of images; Charlotte Amherst and Katie Scott for their illustrations.